Primate Biogeography

Primate Biogeography

Manoranjan Prasad Sinha

Delve
PUBLISHING

www.delvepublishing.com

Primate Biogeography

Manoranjan Prasad Sinha

Delve Publishing
224 Shoreacres Road
Burlington, ON L7L 2H2
Canada
www.delvepublishing.com

Email: orders@arclereducation.com

ISBN: 978-1-77407-610-1 (Hardcover)

ABOUT THE AUTHOR

Prof. Manoranjan Prasad Sinha, M.Sc., (Gold medallist), Ph.D., D.Sc., FAZ, FZSI, FNEA, a Professor of Zoology is currently Vice Chancellor of S.K.M. University, Dumka (Jharkhand) and has served as Pro-Vice Chancellor of V.B.U. Hazaribag. He is a recipient of many awards like Dr. S. Radhakrishanan Education Excellence Award and Scientist of the Century Award, Award of excellence etc. by many learned academic bodies. He is Fellow and Office Bearer of many academic and learned bodies and presently Chairman of Research Advisory Committee of Central Tasar Research and Training Institute and Research Co-ordination Committee members of Central Silk Board Govt. of India. He is credited for publication of more than two hundred research papers in the reputed National and International journals. He is also chief editor of two reputed journals. Prof. Sinha has experience of handling many research projects funded by UGC & DBT, Dept. of Science and Technology, Govt. of India. Prof. Sinha has successfully supervised more than 25 Ph.D. work. The present book is the outcome of some of his findings done with his researcher scholars. He has authored a good number of books.

TABLE OF CONTENTS

GLOSSARY

A

Adaptive Radiation - It is a process in which organisms diversify rapidly from an ancestral species into a multitude of new forms.

Agroecological - the application of ecological principles to agricultural systems and practices

Allogrooming - social grooming between members of the same species

Anthropogenic - caused by humans or their activities

Anthropogenic - chiefly of environmental pollution and pollutants) originating in human activity.

Anthropoids - They are a higher primate, especially an ape or ape-man.

Anthropologists - a person engaged in the practice of anthropology

Arboreal - Living in trees

Arboreal - Covered in woods.

B

Beneath - In a lower place.

Bushy - Growing thickly.

C

Canopy - The uppermost branches of the trees in a forest, forming a more or less continuous layer of foliage

Carnivores - an animal that feeds on other animals

Cathemerality - behaviour of organism in which an organism has sporadic and irregular intervals of activity during the day or night in which food is attained, socializing with other organisms occurs, and any other activities that is required for livelihood are performed.

Continental drift -The shift of continents in place

Correlation - A mutual relationship or connection between two or more things.

D

Deforestation - Cutting of Forest

Dispersal - An act or process in which something scatters or is scattered.

Drudgery - Tedious, menial and exhausting work.

E

Ebola - A Type of Infectious Disease

Electrocuted - Injure or kill (someone) by electric shock.

Endocrinology - It is a branch of biology and medicine which deals with the endocrine system, its diseases, and its specific secretions known as hormones.

Enigmas - A person or thing that is mysterious or difficult to understand.

Eocene Epoch -It is a major division of the geologic timescale lasting from 56 to 33.9 million years ago.

Extant -Still in existence

Extinct - Having no living members of a species

F

Fauna - Animal world.

Feces - Solid matter discharged from a human or an animal's alimentary canal.

Figs - a soft pear-shaped fruit with sweet dark flesh and many small seeds, eaten fresh or dried.

Flora - Vegetable life.

Folivores - An animal that feeds on leaves.

Fossils - Remains and traces of ancient organisms

Fragmented - Break or cause to break into fragments.

Frugivory - A frugivore is an animal that thrives mostly on raw fruits, succulent fruit-like vegetables, roots, shoots, nuts and seeds.

G

Goggling - to stare (at something) with wide eyes.

Granivory - interaction in which seed predators) consume the seeds of plants as a main or exclusive food source

Guild - subset of species inside a community that utilizes a lot of resources in the same manner.

H

Habitat Loss - Process by which natural habitat becomes incapable of supporting its native species.

Herbivores - an animal that feeds on plant

Heterozygosity - the possession of two different alleles of a particular gene or genes by an individual

Hindlimbs - a posterior limb on an animal

I

Inbreeding - Breed from closely related people or animals, especially over many generations.

Infancy - The state or period of babyhood or early childhood.

Infanticide - The unlawful premeditated killing of one human being by another.

Insectivorous - a carnivorous plant or animal that eats insects\

L

Locomotion - The ability to move from place to place, or the act of doing so.

M

Mammalian order - denoting a mammal

Mesozoic Era - It is the age of the dinosaurs and lasted almost 180 million years from approximately 250 to 65 million years ago.

Miocene Epoch - It is the first geological epoch of the Neogene Period and extends from about 23.03 to 5.333 million years ago.

Monogamous: Abstaining from extramarital, or from all sexual intercourse.

N

Natural Selection - It is the differential survival and reproduction of individuals due to differences in phenotype.

Neanderthal - Separate species of humans which became extinct and they are separate from the modern-day humans

Neotropics - A zoogeographical or phytogeographical region comprising Central and South America, including the tropical southern part of Mexico and the Caribbean.

Nocturnal - Happening or done every night.

O

Oligocene Epoch - It is a geologic *epoch* of the Paleogene Period and extends from about 33.9 million to 23 million years before the present.

Omnivores - an animal or person that eats a variety of food of both plant and animal origin

Optimal - Best or most favorable

Optimism - State of having positive beliefs.

Orangutans - long-haired, orangish primates that are highly intelligent and are close relatives of humans.

P

Paleontologist - A person who studies the fossil traces.

Paleotropical region - tropical areas of Africa, Asia and Oceania (excluding Australia and New Zealand)

Phanerozoic Era -2.5 Billions of years ago from the present time

Primatologists- scientific study of primates

Prosimians - They are a group of primates that includes all living and extinct strepsirrhines, as well as the haplorhine tarsiers and their extinct relatives

Pygmy Marmosets - small species of new world monkey native to rainforests of the western Amazon Basin

Q

Quadrupedalism - a form of terrestrial locomotion in animals using four limbs or legs.

Quadrupedalism - It is a form of terrestrial locomotion in animals using four limbs or legs.

S

Scrape - An embarrassing or difficult predicament caused by one's own unwise behavior.

Semi-terrestrial - frequenting but not living wholly on land

Snout - The projecting nose and mouth of an animal, especially a mammal.

Stimulation - A thing that stimulates someone into action.

T

Tectonics -It is the scientific study of the deformation of the rocks that make up the Earth's crust

Topographic - Relating to the arrangement of the physical features of an area.

Transmissions - process of transmitting something

U

Utilitarian - Designed to be useful or practical rather than attractive

V

Vulnerable - the incapability to withstand the impacts of a hostile environment.

W

Watchword - A word or phrase expressing a person's or group's core aim or belief.

Z

Zoochorous - having the spores or seeds dispersed by animals

Zoonotic - diseases spread between animals and people.

LIST OF FIGURES

LIST OF TABLES

PREFACE

Primates have been on this earth for a long time. They have been around, from even before the humans and there have been quite a lot of theories that have also supported the fact that the primates have been the source of evolution of the humans and formed their assumptions around this belief itself.

There have been quite a lot of studies on the primates, which have been conducted to understand the way they live and the various activities they get involved in. Biogeography of primates also aims to know about the various aspects related to the primates so that the researchers can know more about the related things and be able to use them for the betterment of the primates, which can help them in their existence and aid in the various conservatory activities related to the primates.

This book tries to make the readers aware about the biogeography of primates and bring to their knowledge, the various kinds of aspects related to the existence of primates. It starts off by introducing the readers, the meaning of biogeography and informing them about the types of it. The readers are told about the scope and importance of biogeography across various verticals and explained the event of continental drift and tectonic shifts, with their relevance to the evolution of earth.

In the following chapter, the book explains the meaning of primates and informs the readers about their origin. The early evolution of humans has also been explained in brief. Then the chapter shifts the focus to the adaptation in the primates, and to that of structural, behavioral and physiological form. Moving further, the chapters throw light on the classification of the primates and their historical significance and build up.

The next chapter discusses the distribution and diversity of the primates, by bringing to notice, the size range of the primates, natural groups of the extant primates, their distribution and the conditions suitable for their survival. It discusses whether the humans have been descendants of apes or not.

Further, the book informs the readers about the interactions between the primates and other species, discussing the predator-prey interactions and the social structure of the primates and other species. It discusses the interactions of chimpanzees with non-human species and the occurrence of primates as a natural heritage of conflict resolution.

Then, the book discusses about the conservation of primates, throwing light on the reason for their conservation and the threats that are being posed to the primates. The readers are informed about the role of primates and the various conservation policies that are underway. The book also informs the readers about the reasons that make the existence of primates important.

Moving further, the chapters throw light on the challenges that may arise in the conservation of primates and the point of view on their rehabilitation and reintroduction. The readers are informed about the challenges that arise in captive breeding and the ones that have been imposed by time.

In the final part of the book, the readers are informed about the progress of primates and the future of primate conservation, which has been linked to their evolution. The book discusses about the future of primate extinction and lays down directions for primate conservation. The readers are also explained the prospects of genetic modifications in the primates.

This book has been written to make the readers aware about the situation of primates and the various aspects relates to them, so that the readers know about their existence to the best extents and are also updates with the importance of their conservations and the various measures that have been taken to undertake such conservation.

Chapter

1

Introduction to Biogeography

"Every species has come into existence coincident both in time and space with a pre-existing closely allied species."

— Alfred Russel Wallace

CONTENTS

Biogeography is known to be a interdisciplinary field which has a important role in understanding the evolution of different living organisms. There are different types of biogeography that have been explained and the nature, scope and importance of biogeography have been elucidated along with role of biodiversity in evolution, biodiversity and scientific predictions. The influence of the geographical factors such as the continental drift and plate tectonics and the role in evolution have been explained in detail in this chapter.

1.1. INTRODUCTION TO BIOGEOGRAPHY

One common aspect that all of us across the surface of the earth share is that we live on the same planet. Our planet earth is a home for all the living organisms, and this creates a pressure on everyone because there is a constant rise in population and decline in the resources on the surface of the earth. There is an increase in the research on the relationship between the earth and the living organisms on the earth so that there is a chance that the researchers would come up with some suggestions to protect the planet earth. A +major branch that is associated with the relationship between the living organisms and the planet earth is known to be the biogeography.

Biogeography refers to the relationship between living organisms and their distribution over a period of time. Biogeography is one of the most important branches of science and people who work in the field of biogeography try to find answers to some basic questions such as the reason for so many living organisms, reasons for the diversity among these living organisms, the present distribution patterns of these organisms, the future distribution of the living organisms and how the activity of human beings would have an impact on the distribution of these organisms are certain aspects that are studied as a part of biogeography.

Biogeography is a branch which is interdisciplinary and has a lot of importance right from the historic times. The history of the biogeography begins from the ancient times and it is a branch which is a combination of science, geography, paleontology, and ecology.

The increase in travel and several tours across different continents in the 18th century led to the rise of a branch called as biogeography. The distribution of living organisms based on the topology and geography of the earth and the relationship amongst these parameters is known to be the essence of biogeography. The information that is obtained from most of these voyages is kept intact

Biogeography combines several fields of study. It dates back to a long time. Biogeography originated by combining the interconnected fields of geography, biology, ecology and paleontology. There were several geographical voyages in the 18th century. The field of biogeography was born due these voyages.

Till the present times, there continues to be addition to these records. These data contribute to be the basis for biogeography. When Darwin undertook the voyage Beagle, in which he gathered and investigated the biogeographical material. This ultimately paved way to the concept of evolution supported by natural selection. This idea transformed the way biology field evolved. In order to comprehend the conceptual framework of the present-day biogeography, an understanding of how the field evolved over several years is necessary.

The world today is characterized by a huge diversity of organisms. This is perhaps the most remarkable feature of the world today. The bio geographers are facing a major challenge in describing this diversity in detail and also the factors that contribute to the varying patterns of occurrence of the various species on the surface of the planet.

For example, the bio geographers have to explain how more than one species of the same animal or bird is present. How the various species have different patterns of distribution, like some are found only locally, while others are widespread. Why a more species of a certain insect is found in one country compared to another.

There are several such questions that arise and bio geographers not only have to answer these specific questions, they also have to seek general rules that can explain these questions. There are several observations which provides a broad framework which can explain the possible outcomes if the natural world is meddled with.

Edward O. Wilson, an expert on biological diversity asserted that in the time when the humans evolved, there was greater diversity of life on Earth than it was ever present before that time during the entire history of Earth. The emergence and cultural advancement of humans had a significant bearing on the world's biogeography. Even today, this impact continues, and it modifies the species ranges and has even endangered and caused extinction of many species.

From the perspective of biological diversity, it can be said that the evolution of mankind had disastrous consequences for Earth. It would not be realistic to try and combine biogeography without considering the impact of human actions. Almost all species are impacted due to human activity, atleast in some aspect of their ecology and distribution.

Hence, the mankind has played an important role not only because of their influence on other species of animals and plants, but also because men are also one species among many and abide by the same rules which 30 million other species obey.

A better the understanding about the Hawaiian goose, the dodo and the oak tree will help in appreciating man's position in the order of things. Biography draws information from other disciplines because it faces various issues. In order to explain biodiversity, climate patterns must be interpreted in all regions of Earth. It is also important to understand how the plants across various climates and latitude, the photosynthesis of plants varies.

Mankind has to understand why certain habitats are more conducive to plants and animals; why other factors like moisture levels, soil chemistry, temperature variance or spatial structure is more desirable. In order to answer these questions, many disciplines have to be invoked like geology, climatology, soil science, ecology, physiology, and behavioral sciences.

The field of biogeography is one that throws light on the various kinds of disciplines that focus on the study of various types of species or organisms. These may involve various disciplines such as

- Pedology
- Geology
- Planetology and
- Climatology

These are studied in a proper way and the study of these disciplines may vary depending on the differences in their scales and time of interest and this may also concern the number of species and the nature they may be having.

This may be illustrated in a way that some of the organisms may be occurring at a very small scale and this may lead to a requirement of transforming their numbers into bigger amounts.

The concepts regarding the biogeography don't require to be put away and are very applicable. For biogeography, various theories have been made, that may involve the Equilibrium theory of island biogeography. This theory focuses on the rate of extinction and the ways in which species migrate, along with a focus on equilibrium rate. This makes some provisions for the proper conservation and management of the resources.

Biogeography helps the researchers to understand the various processes that the living organisms go through and the various patterns related to it. It aids them to know better about the various trends related to those organisms and the processes they may have undergone.

The field of biogeography and the undertaking of the related study makes it easy for the scholars to know about some species and also get to know about the plants and organisms in a particular place or a different place.

The study of biogeography also helps to know why the flora and fauna at a particular place are different from the ones in some other place, which may be happening due to some critical factors in the environment and also bring to notice the manner in which various ecological and historical factors might be influencing a thing which may be taking place in a certain way. For instance, there are certain environmental species that are tolerant to the environment and ones that survive and reproduce in the same environment.

The field also helps in understanding the impact of the activities undertaken by the humans, on the ecosystem and the environment. In such environments, the concerned elements may be expected to act with more coherency and consistency, as they may be required to handle the activities that may be taking place across various verticals.

For instance, some of the application including the conservation of fisheries and the measures related to their management take up the assessment of stocks and their monitoring and include the conservation of biodiversity and the various bases for incorporating the information involving scientific aspects and the socio-economic aspects related to the usage done by the humans. It may also involve the various possible threats, legislations and the policies related to them.

The Criticism of Biogeography: Limitations and Weaknesses

The scarcity of information regarding the specific way in which extinction and migration of the species happened and the number of species, which may have been there at the point of equilibrium, may impact the geological time. This can especially take place the migration of species and their extinction coincide with the climate and geological events that might have been strange.

In addition to this, the species may differ in terms of the rates of extinction and migration. For example, the rate of migration of birds is generally much higher than the reptiles which may be big in size and heavy and various other mammals. Due to this, the conservation and management of the natural resources becomes difficult.

Moreover, most of the people in the local areas have no understanding of the biogeography concerned with the management and conservation of natural resources. Hence it becomes quite difficult for the people in the local areas to preserve the resources and manage them in a sustainable way. These resources may involve various species and their habitats, which need to be protected from disappearing from the particular region.

Hence, biogeography cannot be taken as a useful tool for the local people, for the purpose of conservation and management of natural resources such as water bodies, land, forest and the various species belonging to various ecosystems.

The theory relating to continental drift, has been quite heavily criticized by some of the researchers and theorists. For instance, they state that the jig-saw fit is not capable of fitting in all the continents properly. In addition to this, there are some of the plants and organisms that have been believed to move from one place to another by the action of wind, which further weakens the theory.

1.1.1. Biogeography Definition

Biogeography is known to be a field of science that deals with the distribution of several species of organisms across the different types of ecosystems that are spread at different geographical locations over a period of time and space.

Biogeography is known to be studied as a branch of science in the context of ecological and historical factors that would be take form over a period of time across different topographical locations. The study of biogeography is often carried out in the context of ecological and historical factors that have shaped the geographical distribution of organisms over a period of time.

There are different species of organisms that are present at different geographical locations based on the latitude, habitat, segregation, and elevation of the geographical location. Biogeography can be divided into several sub disciplines such as zoogeography and phytogeography. Zoo geography refers to the field of biology that deals with the distribution of living organisms such as animals based, and phytogeography refers to the distribution of the living organisms such as plants.

1.2. TYPES OF BIOGEOGRAPHY

There are three important categories in the field of biogeography, and they are as follows:

- Historical Biogeography
- Ecological Biogeography
- Conservation Biogeography

These three types of biogeographies would address the distribution of the various species of organisms from a different perspective.

1.2.1. Historical Biogeography

The historical biogeography refers to the distribution of organisms over a period of time across a specific area based on the phylogenetic characteristics and evolutionary significance.

1.2.2. Ecological Biogeography

Ecological biogeography is a branch of biogeography that comprises of the relationship between different factors of the environment that contribute to the distribution of various plant and animal species on the distribution of living organisms. There are different types of ecological factors that would influence the distribution of living organisms and some of the important ecological factors are climate, primary productivity, habitat and several other similar factors. The rate at which the plants or animals that are living in a particular ecosystem generate the net chemical energy is being studied.

The ecological biogeography of several living organisms is known to be a lot different from the historical biogeography of the living organisms and the rate at which the distribution of these organisms vary over a period of time is known to be completely different.

1.2.3. Conservation Biogeography

There is a concept of conservation biogeography is known to be the process by which the current level of biodiversity is managed throughout the word by providing different policy makers with data and potential concerns related to the conservation biology.

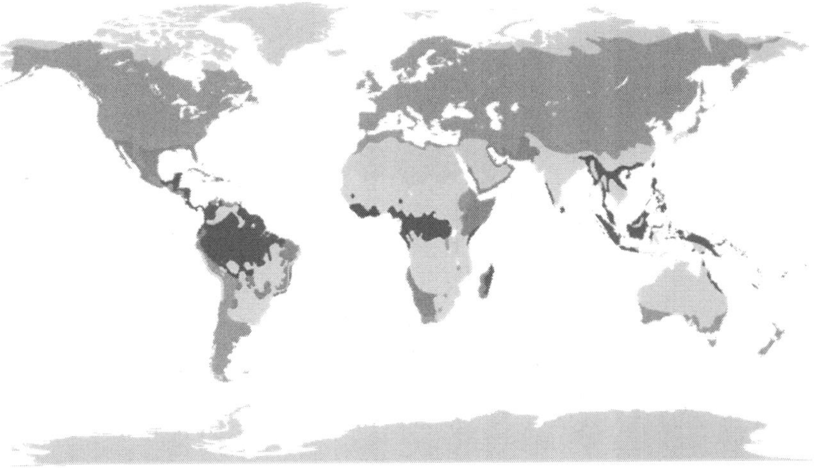

Figure 1.1. Representation of Biogeography.

Source: Image by wikimedia

1.3. NATURE, SCOPE AND IMPORTANCE OF BIOGEOGRAPHY

Biogeography refers to the study and distribution of the biodiversity of several living organisms over a period of time and space. The major objective of the biogeography is to identify the places where the living organisms live in excess and the reasons for the presence of living organisms in higher numbers at a particular region.

Biogeography can be present at both regional level and global level at the international level biogeography refers to the science that involves the study of the distribution of several continents and intercontinental connections that are present between continents to describe the variations that are present between the plants and animals that are present in that particular continent. At the regional level, the study of the distribution of different species and communities according to the topography and the environmental conditions along with the geographic and climatic barriers that bring about the differentiation between several species and population that is present in the region. At the local level, biogeography refers to eh study of the distribution of several species and different living organisms that are present in the microhabitats. There are two major branches of biogeography and they are as follows:

- Phytogeography
- Zoogeography

1.3.1. Phytogeography

Phytogeography refers to the branch that deals with the distribution of several species and communities of plants that are distributed across several areas.

1.3.2. Zoogeography

Zoogeography refers to the branch of biogeography that refers to the abundance of animals that are present over a space during a particular time.

Natural resources refer to the untouched resources that are offered by the earth naturally and are not disturbed by the human activity in any form. The natural resources are known to be valuable and are useful to man in their original natural form depending on the quantity and demand of these resources. Some of the important natural resources are land, forest, energy sources and minerals.

Natural resource management is a subject that refers to the organization of several natural resources such as land, soil, water, plants, and animals which have a specific attention on how the handling and management of resources affects the quality of the life of both present and future generations.

Natural resource management mostly deals with the management of the methods by which the natural landforms and people interact with one another. Natural resource management comprises of different aspects such as the planning the utilization of land, Management of different water resources, conservation, and the sustainability of the industries such as agriculture, fisheries, tourism, and forestry.

Conservation of natural resources generally refers to the optimum utilization of the natural resources that are provided by the earth. Human beings need to play an effective role in conservation and management of earth resources. Conservation of the natural resources on the surface of the earth refers to the undistributed utilization of the resources in such a way that these resources are mostly capable of self-renewal and generate few nonrenewable resources such as coal and oil.

Biogeography is known to be an important tool that would lead with the people who conserve and manage natural resources to understand the interaction between different organisms that compete for food. The interaction of the living organisms and their interaction with each other while competing for food is known to be an important aspect while studying the biogeography of living organisms. The relationship between the environment, living organisms and their eating habits are the important aspects that determine the distribution and abundance of the living organisms.

The interactions that are present among several living organisms are extremely important and they sometimes get so complicated that the variation in the abundance of one animal or plant might have several changes in the other organisms and may lead to a lot of changes in the entire community. Therefore, it is important to understand that the ecologists and bio geographers must have knowledge regarding the changes in the relationships of the organisms that occur and the result of the environmental change or adoption of the living organisms to these changes.

Biogeography is known to be an extremely important tool as it facilitates researchers in understanding of different climatic patterns that are present on the surface of the earth and this field is known to be related to the field of climatology that deals with the understanding the relationship between the climatic changes and the influence of these changes on the distribution of living organisms and help in predicting the changes that would occur in these organism in the future.

Biogeography is known to be an essential tool that helps in understanding the various fossil records that are learnt through various aspects of historical biogeography and they facilitate looking into the past and observe the different changes in diversity and the distribution patterns of living organisms with time. These regulations and rules are known to have played an important role in the conservation of natural resources in an appropriate way.

Biogeography is known to have been used as a base for research, forecasting and proactive management of the natural resources that involve several ecosystem responses that are related to climate change. The examples of these mostly include the prediction of the area of habitats or species of the organisms

Biogeography generally refers to a branch of study that deals with the study of several areas of science that deal with the study of living organisms such as the pedology, paleontology, geology, climatology and several other areas which vary in their depth of knowledge and change at several scales and times of interest with respect to the number of species of living organisms and their nature.

Biogeography refers to the study of the distribution of different life forms at different geographical areas over a period of time. Biogeography can be broadly divided into two different categories called as the ecological biogeography and historical biogeography. Ecological biogeography deals with the different types of distribution patterns for various living organisms and the historical biogeography deals with the long-term and large-scale distribution patterns of several living organism.

1.3.3. Biogeography and Biodiversity

Biogeography is known to be a field of science which is a combination of the concepts of biology and geology. There are some eminent biogeographers such as C. Barry Cox and Peter D. who write about biogeography as a field of biology that explains the patterns of distribution and the interaction of species. They state that biogeography is a field which deals with the relationship between evolution of living organisms and the plate tectonics of several organisms,

The concepts of biodiversity may be explained in relationship with the climate and latitude. The studies indicate that there is need for various organisms to live in different habitats such as animals and plants and the reasons for the presence of a specific soil chemistry, moisture levels, temperature range or spatial structure of different organisms must be extremely interesting. Therefore, the different branches of the science such as climatology, soil science, geology, physiology, ecology, and behavioral sciences are being explored to find answers to the distribution of these living organism. Biogeography is known to be extremely important in the analysis and distribution of different organisms. These studies facilitate in understanding of the various changes in distribution that would arise and the factors that contribute to the changes in the distribution of living organisms and their adaptations.

1.3.4. Biogeography and Scientific Predictions

The basic capability of science is to make predictions based on the theory that is available. The extent to which these predictions are successful to explain the concepts of evolution is still a question. The scientific predictions in biogeography are known to have played an important role in theory of evolution. The reasons for some organisms to have a greater mobility when compared to the other such as sea animals, birds, animals, humans are known to have occurred because of a concept called as plate tectonics and there are several concepts of evolution that have been predicted and explained by the concept of biogeography.

The distribution of species is known to have happened in a random manner depending on the geographical barriers that are present on the surface of the earth. There are some species which are closely related to one another and the other species are known to have a of diversity the reasons for the similarities and variations in these species was explained with the help of biogeography. Biogeography is a concept that can be well explained by evolution and the evolution of the organisms and the plate tectonics are capable of making predictions about the similarities and dissimilarities amongst these organisms.

1.3.5. Biogeography and Evolution

The biogeographic distribution of living organisms belonging to different species is known to support the concept of evolution. There are a number of species of living organisms that have been distributed across the globe and these organisms are known to be different from one another and at the same time they are known to have a genetic relationship with one another. The climatic conditions and environmental factors of a place are known to have an impact on the adaptations that are made by the organisms and the changes that happen in different organisms. For instance, marsupials are known to be found in Australia and these organisms have been restricted from the other parts of the world and the presence of placental mammals is known to be extremely rare in Australia. The presence of marsupials across the world would have made it extremely difficult to explain the concept of evolution in terms of biogeography.

There are only few placental mammals which are naturally native to Australia such as the murine rodents, seals, bats, sealions, dugong, whales, dolphins, and seals. Bats are known to have flews to Australia nearly 50 million years ago, sea mammals are capable of swimming and they would have reached Australia by swimming. Some animals such as murine rodents are known to have Australia because of the arrival of several human colonists especially 50,000 years and 200 years ago. There are some groups of mice and rats called as the "old endemics" are known to have been found as fossils nearly 4 million years back in Australia. Murines are known to be some of the most common mammals that are present on the surface of the planet and are considered as adept at the dispersal of the oceans by events known as rafting events.

1.4. CONTINENTAL DRIFT AND PLATE TECTONICS

Researcher Wallace is known to have accepted the theory of evolution by the end of his voyages to amazon and Southeast Asia in 1848.These journeys made by Wallace have led to the discovery of the influence of the geography on the evolution of different living organisms. Wallace tried to demonstrate the theory of evolution and the influence of the geography on the evolution of different organisms. During the voyages made by Wallace there were several species of plants and animals that were studied and the inference that was obtained from these organisms is known to have explained the theory of evolution.

The patterns of evolution that were identified by Wallace are known to have a lot of information that would provide an evidence for evolution. There were several aspects that were explored by Wallace such as the reason for the formation of different rivers and mountain ranges and the influence geographical barriers on the evolution of different living organisms were explained. The theory that was present by then stated that there was diversity that is present amongst different living organisms because of the adaptations that were made by different organisms to the changes in climate. However, Wallace did not find this theory satisfactory as there were different living organisms belonging to various species that were present in the similar climatic conditions.

The studies conducted by Wallace led to the same conclusion that was published by Darwin in the Origin of species which states that biogeography is majorly a simple record of inheritance. There were different species that were in different environmental conditions and there were different animals that occupied diverse mountain ranges and other places which led to the diversity in the distribution of species.

There are different types of biogeographic regions that are present across the world and Wallace stated that these continents coincide with one another. The recent studies about the biogeography reveal that it has a dynamic role in the evolution of life on earth over a period of time. There are several interesting discoveries made by scientists in relation to the importance of biogeography and one of these important discoveries is that a geologist named Alfred Wegener identified similar fossils of plants and animals on either side of the Atlantic Ocean.

This identification revealed the fact that because it was extremely difficult for the animals or plants to cross the ocean on their own the identification of similar fossils on either side of the Atlantic Ocean reveal the fact that these both continents were once connected to each other. The scientists and geologists could carefully map the ocean floor by the end of 1960 because of which the concept of continental drift of ocean was possible because of a mechanism called as plate tectonics.

Biogeographers have identified the fact that there are several continents that would collide with one another and there would an interaction among the species and when the continental parts separate then there would be new species that would grow in the continents. The studies reveal that there was a time at which all the continents such as Africa, South America, New Zealand, and Australia were joined with one another to form a super continent. This super continent was popularly called as Gondwanaland.

These continents then started drifting apart from one another starting with Africa, followed by New Zealand and then Australia and South America. The phylogenetic analysis of certain living organisms such as midges indicate a similar evolutionary pattern. For instance, certain South America and Australian midges are known to be closely related to one another and they are comparatively related less to the midges in New Zealand and the midges that belong to all these three areas have less relationship with the midges that are living in Africa. This indicates that a small insect which has a lifespan of about few days to weeks is capable of giving evidence to the continental drift and biogeography of a place that has happened several million years back.

The theory of continental drift provides an important relationship between the plate tectonics and biogeography of a place. The formation of the continents by continental drifts is explained below to facilitate with the understanding of biogeography of the earth.

The beginning of the earth which happened nearly 4.6 billion years ago indicates that the earth was a ball of burning gas which was spinning through space. At the beginning these super-heated gases were capable of escaping into the outer most space and as the temperature on the earth cooled down the gravitational force on the surface of the earth could hold the atmosphere. The arrival of the atmosphere on the surface of the earth is known to lead to the arrival of life on the surface of the earth.

The presence of atmosphere led to the development of water vapor which was getting collected in the air and this water vapor that was collected was falling down as rain, which led to the rising of the ocean. There was lot of time duration nearly few millions of years that took for the arrival of the first land mass on the surface of the earth.

Continental drift is one of the oldest theories by which geologists explain the movement of continents over a period of time. This theory of continental drift was replaced by the concept of plate tectonics in the recent times. The concept of the continental drift is mostly associated with a scientist called a Alfred Wegener and this scientist is known to have published a paper in the 20th century to explain the concept of the movement of land masses and how they "drift" across the earth across the oceans and collide with one another. Alfred Wegener has called this process of the movement of the landforms as continental drift.

1.4.1. Pangaea and Gondwana

The first super continent called as the Pangaea was formed nearly 250 million years back which is a long way after the Earth was formed by the merger of all the landforms. This super continent called as the Pangaea separated to form two giant continents called as the Gondwana and Laurasia. Alfred Wegener in his research supported the concept of the existence of one huge continent called as Pangaea which has split into two smaller continents called as Gondwana and Laurasia.

Gondwana was a super continent that comprised of several of the recent continents such as Africa, Antarctica, India, South America, and Antarctica. The Indian Subcontinent was present at the east coast of Africa and later it separated and moved North and collided with Asia. This collision of Indian Subcontinent with Asia led to the formation of the largest and greatest mountain ranges on the surface of earth called as the Himalayas. The great Himalayan mountain range extended for a range of about 2500 kilometers.

Alfred Wegener was trained in the field of Astronomy and he used the concepts of biology, botany, and geology to describe the concept of the continental drift and the formation of the Pangaea super continent. For instance, there were fossils of some ancient reptiles such as Mesosaurus which were found in the continents of southern Africa and South America. Mesosaurus was a freshwater reptile which was nearly one meter long (approximately 3.3 feet) and it is impossible for this small reptile to have swim across the Atlantic Ocean. The Mesosaurus reptile is known to have a single habitat and would be present in many lakes and rivers.

The studies by Wegner suggest that there are different types of plant fossils that were identified from the Arctic Archipelago of Svalbard, Norway and these plant specimens were showing adaptations to the warm and humid climatic conditions of Svalbard, Norway. These plants suggest that Svalbard, Norway initially had a tropical climate and later there was a change in the climate.

The final studies of Wegener related to the biogeography comprised of the study of the different rocks and mountain ranges that are present on the earth. This study is generally called as the stratigraphy of various rocks and mountain ranges. The studies indicate that there are some places such as the east coast of South America and the west coast of Africa which were fitting into one another like pieces of a jigsaw puzzle and Wegener discovered that these rock layers were clearly fitting into one another. In addition, the research and studies reveal that South America and Africa were not the only continents which has similar geology there were several countries such as Appalachian Mountains which were a part of eastern United States were known to be geologically related to Caledonian Mountains of Scotland.

Pangaea is known to be an island that existed nearly 240 million years ago and nearly after 200 million years this supercontinent started breaking away from one another. These pieces gradually started assembling from one another to form those continents that are similar to those that we recognize today.

The recent studies indicate that there ae several supercontinents such as Pangaea which were formed and separated from one another over a period of time during the course of evolution of Earth. Some such continents that have formed and separated in the recent times are Pannotia a super continent which was formed nearly 600 million years back and Rodinia which was a continent that was formed nearly a billion years back.

1.4.2. Tectonic Activity

There were several scientists and researcher who did not approve the Wegener's theory of continental drift as the major aspect that was missing in the Wegener's theory of continental drift is the mechanism by which the process of continental drift work occurs. The mechanism that was suggested by Wegener for this continental drift is that "earth's rotation could be the reason for the drift of these continents", but the recent studies suggest that these continents

The studies and research provide with the knowledge that the continents rest on few massive slabs of rock that are called as the tectonic plates. These tectonic plates are known to continuously moving and interacting with one another by a process called as plate tectonics.

The studies and the research on the continental drifting suggests that these continents are continuously moving even in the present times and some of the major places where there is dynamic tectonic activity is the seafloor spreading zones and the giant rift valleys.

The process of seafloor spreading refers to the process where the molten rock rises from the internal regions of the earth and adds the new seafloor or oceanic crust to the edges of the old. The process of seafloor spreading is known to be extremely active along the giant underwater mountain ranges which are called as the mid-ocean ridges. The increase in the area of the seafloor would push the continents on either sides and continents on the opposite sides are known to move away from one another.

The North American and Eurasian tectonic plates are known to be separated from one another by the Mid-Atlantic ridge. These two continents are known to be some of the most important continents that are known to be moving away from one another at the rate of nearly 2.5 centimeters (approximately 1 inch) per year.

Rift valleys are known to be those sites where the continental landmass was ripping apart from one another. For instance, some continents such as Africa are known to have separated from the landmass at the Great Rift Valley System. This separation at the Great Rift Valley system is known to have emerged into two continents one on the African Plate and the other one on the Somali plate. The new continents on the Somali side were known to oceanic continents with the Madagascar and the Horn of Africa as the biggest land masses.

The procedures which involve seafloor spreading, rift valley formation and subduction are known to involve the sinking of heavier tectonic plates beneath the lighter plates. This procedure was not completely established before the 1960s and later than these concepts were well formulated. These processes are known to be the important mechanisms that are known to be responsible for the concept of continental drift that was explained by Wegener.

The mechanism of plate tectonics led to the formation of several land masses on the surface of the earth is responsible for the separation of the land masses and formation of continents. This process of plate tectonics is still continuing and the movement across the earth's crust is further continuing. The studies and research by several scientists indicate that these small continents would converge with each other to form a new super continent.

1.4.3. Relationship between Biogeography and Evolution

Biogeography is known to be an important field of biology as it gives evidence of evolution of different types of living organisms that are present across different landforms and diverse environments on the surface of the earth. In due course of time the Earth's continents are known to have separated, drifted apart and collided with each other which led to the formation of different types of climates and habitats. The different species of organisms started getting adapted to diverse environmental conditions and the living organisms that were belonging to the same species are known to be geographically diverse and this led to the formation of a unique species of organisms. There is need to study intensely and conduct a lot of research to understand the reasons that have led to the formation of different adaptations and the reasons by which these organisms adapt differently to the same climatic conditions is known to be understood and this knowledge must be applied in the future research.

1.4.4. The Galapagos Islands as an example to Biogeography

The major example of biodiversity in relation to the evolution of living organisms is provided by Charles Darwin's study of the finches that are present on the Galapagos Islands. These studies made by Charles Darwin led to the arrival of a book called as Origin of Species.

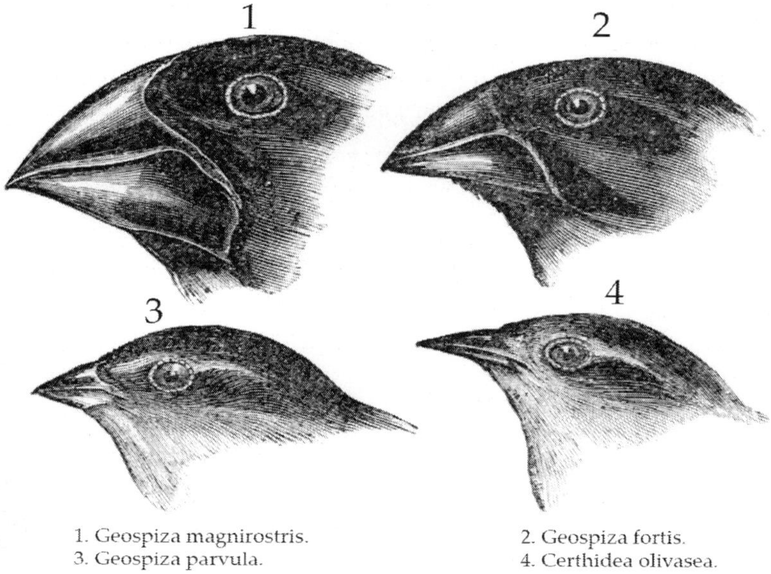

Figure 1.2. Representation of Darwin's Finches.

Source: Image by wikimedia

The research of Darwin on the finches living on the Galapagos island led to some interesting discoveries that suggested at the finches which were living on the mainland of the South America were known to be similar to those finches which were living on the Galapagos Islands. However, there were some minor differences in the physical appearances of these finches such as the shape of the bills of finches were known to be different. The reason for the difference in the shape of these bills were known to be because of the availability of different food choices on these Galapagos Islands.

The Galapagos is lands were known to be a part of the South American mainland initially and with the due course of time these islands are known to have drifted apart from the South American mainland. This separation of the islands form the South American mainland led to the creating of different new habitats and food sources which are available for the several species that are residing within these regions.

This suggests that each finch species is known to be adapted to the local environment through the selection of those alleles that would prompt the survival of several species.

Islands are known to be extremely important in the study of the biogeography of the small ecosystems and the information that is obtained would be immediately compared to the animals and their adaptations that are present in the mainland. The islands are known to be isolated landforms which would have different invasive species that are associated with different types of consequences for different types of organisms that are present within the ecosystem. The study of these changes and adaptations that happen in different living organisms would give a greater knowledge about the species and the ecosystems that are present in different living organisms.

1.5. BIOGEOGRAPHY AND RELATIONSHIP TO EVOLUTION

Charles Darwin (1809-82) is famously known for its theory of evolution. He overemphasizes the importance for natural selection – that over time creatures, which possess the ability to adapt biologically to changes in their environments (in other words, evolve) survive the prevailing climatic conditions, while those that don't able to cope with the uncertainty become extinct.

He also made his point that all species of life on Earth are interrelated with each other and includes a common ancestry, which belongs to an era when the life started. Evidence that are weighing this theory can be viewed in hominid fossils from the Sterkfontein Caves as well as from other sites in the Cradle of Humankind. These fossils help in explaining that our early hominid ancestors had human-like teeth and had the ability to walk on two legs, but that they also comprised of different ape-like features, including small brains.

Many scientists have the opinion that hominids separated from the ape lineage and this occurred almost around 8–million and 7–million years ago.

It's has been only within the past 50 years or so, that the theory of evolution has received acceptance from all over the world. But the Western world had learnt about the presence of human-like animals, the great apes, in Asia and Africa, by the 18th century.

Carl Linnaeus, a Swedish botanist who was recognized for developing a scientific system with which to characterize and classify all living things in 1735, decided that both humans as well as apes possess certain similar characteristics that are enough to classify them collectively in the zoological order primates.

Forty years later, James Burnett, Lord Monboddo of Scotland, who was an eccentric judge and philosopher, explained that humans were closely associated to orangutans, and that Africa was belonged to the humanity's ancestral home. Burnet is now accolade with being the very first scholar that introduced the concept of evolution.

In 1859, Charles Darwin published the widely recognized book, theory of the evolution of species, and explained that humans had evolved from older, yet to be discovered species.

The first fossil that was found to be of a human ancestor was a Neanderthal skullcap, that was discovered in the year 1856 in Germany.

In the same year, a fossil of an early age, Dryopithecus, was also discovered in France. One of the ancient types of human, Pithecanthropus, later reclassified as Homo erectus and now primarily known as "Java Man", was found in the year 1891 in Java, Indonesia.

In the year 1924, the skull of the most ape-like human ancestor yet, the Taung Child, was found in what is now known by the name the North West Province of South Africa. It had brain which is of ape-sized, but teeth which is similar to humans.

Professor Raymond Dart, the anatomist who regarded it as a hominid, claimed that this fossil signifies a relation between apes and humans and named it as *Australopithecus af ricanus* ("southern ape of Africa").

In the year 1936 paleontologist Dr Robert Broom discovered the first adult *Australopithecus* at Sterkfontein. It is usually seen that hominids evolved from an ape during the time period, known as the Miocene Epoch, which occurred almost 7-million years ago, as the oldest known hominid, which was approximately

7-million-year-old *Sahelanthropus tchadensis* from Chad. Some of the other early known hominids are almost 6-million-year-old *Orrorin tugenensis* from Kenya and the 5.8-million-year-old Ardipithecus ramidus kadabba from Ethiopia.

1.5.1. Principles of Darwin

The biographic distribution of living organisms was considered by Darwin as an evidence for evolution. There are three important principles that support the evolution of the organisms based on the biogeography. Some of the important principles that were suggested by Charles Darwin in the course of evolution were as follows:

One important principle is that environment can neither be accountable for the similarity or dissimilarity of organisms because it is evident that presence of the same climatic conditions and environments can harbor the presence of different species of organisms.

The affinity (similarity) among the different species or groups of these organisms on the same continent is known to be more when compared to the living organisms that are present between several continents.

Geographical barriers are known to have an important role in dividing or separating different groups of organisms and there is known to be a huge correlation amongst the degree of difference and rate of migration or ability to disperse between the geological barriers.

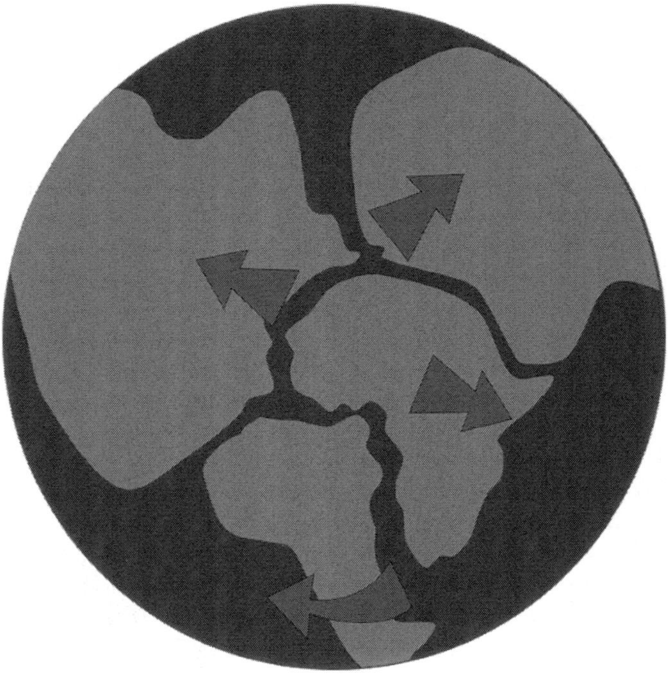

Figure 1.3. Continental Drift

Source: Image by Pixabay

1.5.2. Disjunct Locations for the Same Extant Species

The study of different organisms is known to provide ample scope for the creation and modification of living organisms. There is need to explain the distributions of the living organisms and the Darwin's theory of evolution suggests that there are single centers for the origin of the species, and this is mostly explained by the concept of discontinuous distribution.

The means of dispersal of distribution of the living organisms are proposed as follows:

- The changes in the climate or the geology are known to have a significant impact on the migration. The first effect could be to allow migration and the second could be to prevent the scope of migration.
- Darwin is known to have designed several tests for the priori assumptions
- Though the dispersal of the organisms may be "accidental" this dispersal could not eb random and this facilitates the specific prediction about distribution of organism in certain cases

1.5.3. The Development of Life on Earth

The research and studies indicate that the life on earth initially evolved nearly 3.8 billion years back and ever since life evolved on the surface on the earth the pattern of life in the planet is known to have become increasingly difficult.

Life on earth is known to have developed small simple and small living organisms and from these small organisms there were nearly 1.3 million species of living organisms have evolved. The scientists estimate that there are nearly 30 million to 100 million species son the surface of the earth and only 1.3 million species on the surface of the earth have been documented.

The development of life on earth has happened sturdily and gradually over a period of time, however, there were huge difference in the changes that were happening on the surface of the earth. There were a majority of notable changes that were happening on the surface of the earth in the last eight period of the history of the earth which was called as the Phanerozoic Era which comprises of the last 543 million years on earth.

The first organisms that have evolved on the surface of the earth are known to be the cyanobacteria which are extremely small, single celled organisms that could form a film over a surface of earth, and they were forming layered structures called as stromatolites. Stromatolites are known to be those organisms that have emerged on the surface of the earth after the surface of the earth started cooling down and there was atmosphere and oceans that were formed on the surface of the earth. These small living organisms are known to have become almost extinct, but they are still forming mats in several places that have high saline waters such as Shark Bay in Western Australia.

There are some fossil stromatolites that have been discovered in a wide range of environments such as thermal springs, lakes, seas and certain lakes that are covered with ice in Antarctica. The studies and research conducted by scientists are known to have led to the identification of several fossil traces of different stromatolites which are nearly 3.5 billion years old. These fossils were found near a place called as Barberton which is present in the Mpumalanga Province of South Africa. There were some fossils that were similar in age that were found in north Western Australia and Greenland. There were some fossils of the stromatolites that were found at Sterkfontein. The fossils of the stromatolites are known to be one of the oldest records of life on the planet.

1.5.4. Diversity of Life on the Earth

The studies and research suggest that there was explosion of life that was seen in the last eight of the history of the earth. There were some living organisms such as sponges, jelly fish, flatworms and others that were found on the surface of the pecan nearly 600 million years ago. The first arthropods that were present on the surface of the earth are known to be millipedes, centipedes, spider and scorpions and these creatures are known to have grown onto the surface of the earth nearly 450 million years back.

Insects are known to have evolved nearly 400 million years back and the reptiles evolved about 330 million years ago. The history suggests that the first mammals have evolved nearly 220 million years ago, and the first birds have evolved nearly 150 million years ago.

The first flowering plants on the surface of the earth are known to have come into existence nearly 1118 million years ago. In the due course of evolution there were some species that became extinct and among all the living organisms the dinosaurs are known to have become extinct nearly 65 million years ago and the first primates which are considered as the ancestors to the human beings are known to have evolved nearly 55 million years ago.

The present studies indicate that all the living organisms on the surface of the earth comprise of nearly 4000 different mammals, 9000 birds and nearly 7,50,000 different categories of insects. Though there is a constant arrival of novel living organisms by evolution it must be noted that there are several thousands of living organisms that are getting extinct every year. There are some studies that indicate that there are nearly 100 species of organisms that are dying out every day and the records related to the conservation of living organisms indicate that there are 500 organisms that become extinct every year.

The studies indicate that the only continent which is a remnant of the diversity of the earth is known to be Africa and the presence of extremely limited species of human beings in Africa has left the place completely untouched and led to the conservation of the different species of organisms. There is a rapid change that is happening on the surface of the earth because of human activity which is leading to the loss of biodiversity in living organisms that are present on the surface of the earth. This is leading to a threat to the biodiversity on the surface of the earth.

REFERENCES

1. https://biomed.brown.edu/Courses/BIO48/29.Biogeography.HTML [Accessed 27 Aug. 2019].
2. https://www.thoughtco.com/what-is-biogeography-1435311 [Accessed 27 Aug. 2019].
3. https://bio.libretexts.org/Bookshelves/Ecology/Book%3A_Biodiversity_(Bynum)/13%3A_Biogeographic_Diversity [Accessed 27 Aug. 2019].
4. https://www.learnreligions.com/how-biogeography-supports-evolution-249883 [Accessed 27 Aug. 2019].
5. https://theconversation.com/explainer-what-is-biogeography-10696 [Accessed 27 Aug. 2019].
6. https://www.annualreviews.org/doi/10.1146/annurev.es.26.110195.002105 [Accessed 27 Aug. 2019].
7. https://www.maropeng.co.za/content/page/introduction-to-evolution [Accessed 27 Aug. 2019].
8. https://www.maropeng.co.za/content/page/the-development-of-life-on-earth [Accessed 27 Aug. 2019].
9. https://scielo.conicyt.cl/pdf/gayana/v72n1/art12.pdf [Accessed 27 Aug. 2019].
10. http://www.nyu.edu/projects/fitch/courses/evolution/html/geographic_distribution.html [Accessed 27 Aug. 2019].
11. https://link.springer.com/article/10.1007/s12052-012-0421-2 [Accessed 27 Aug. 2019].
12. Available at: http://socialscience84.blogspot.com/2017/04/nature-scope-and-importance-of.html [Accessed 27 Aug. 2019].
13. https://www.nationalgeographic.org/encyclopedia/continental-drift/ [Accessed 27 Aug. 2019].

Chapter 2 History and Evolution of Primates

"Evolution is one of the half-dozen shattering ideas that science has developed to overturn past hopes and assumptions, and to enlighten our current thoughts."

–Stephen Jay Gould, 1941 – 2002

CONTENTS

Evolution is a gradual process in which the traits of a species transforms over several generations. This chapter explains the concept of evolution. It then discuss what primates mean. The chapter then discusses in detail, how primates evolved. The chapter also discusses the evolution of humans who are the most intelligent primates on Earth. Adaptation is a phenomenon that happens through evolution. This chapter discusses the three types of adaptation in primates, namely structural, behavioral and physiological.

2.1. INTRODUCTION

Evolution refers to the gradual process in which the characteristics of a species changes over several generations. It generally evolves into a different form which is generally environment. more complex or better form. This generally happens as a response to the change in their environment.

Charles Darwin's theory of evolution is the most popular concept which is used to explain how life reached its current state. Evolution is a biological process which is driven by natural selection. Many scientists have used this theory to explain evolution to the extent that it is generally assumed as factual in most researches.

However, the concept of evolution is not absolutely clear. A thorough analysis of evolution raises doubts, the answers of which are hard to find. Evolution has not been proved fundamentally and most experts agree on this.

Ironically, many evolutionists support this theory and fill un the gaps in the evolutionary models using reasonable assumptions. The non-evolutionists are often criticized for not making these assumptions and accepting the theory.

Evolution has many components namely Natural Selection, Micro-evolution and Macro-evolution. The theory of evolution has been around even before any evidence for its components were recorded. Scientific studies eventually were able to interpret data based on the concepts of evolution. They were integrated into the modern theory of evolution.

Evolution occurs through the process of natural selection. Evolution causes organisms which cannot survive in a particular environmental condition to adapt the changes. The Theory of Evolution can cause organism to evolve into a totally different form of life. There is some evidence of natural selection is evident in nature. However, it is not as profound so as to change species in a meaningful way.

Science has observed many changes in the form and function of an organism which has led to death or handicap. This implies that if an organism cannot adapt to changes around them, they will eventually die.

The micro-evolution is widely accepted. Experts believe that coyotes, foxes, wolves, dingoes, jackals and many other domestic dog breeds have all evolved from an original pair of dogs. This denotes variation within a kind. This is not upward evolution from simplicity to complexity which was proposed by the Evolutionary Theory of Darwin.

The variations are always in a downward trend. The genetic code is a constraining factor. For example, a dog cannot grow wings and learn to fly. In evolution of the family of dogs, information was always lost with nothing new being added. While the original pair of dogs would have possessed all the potential characteristics of their progeny, the descendants do not possess the same potential.

The dog kind has many progenies which have become isolated gene pools. The original parents would have possessed all the potentials traits expressed in their progenies. This has been observed by scientists. They have not come across a process which aids in adding genetic information during the process of evolution.

Micro-evolution has been observed and documented well, while the same is not true about Macro-evolution. This makes Macro-Evolution highly disputable. Micro-evolution refers to transition from one kind of animal into another, whereas macro-evolution refers to major changes in the basic function of an organism.

Macro-evolution takes place over a series of genetic mutation and cannot take place during the lifetime of a single organism. Every genetic mutation involving form or function observed in laboratories has either been fatal, crippling, or self-reversing. For example, some bacteria who genetically deviate from their ancestors can withstand antibiotics better than those without genetic deviation. However, they could not survive once the antibiotics were withdrawn.

The evolutionists explain the existence of billions of diverse species on Earth. In other words, how variation from one kind to another occurred.

Evolution as the core mechanism of biology has several shortcomings. It is seen that the organisms suffer more and benefit less from mutation. It is very difficult to explain the balance of Earth's ecosystem and their relationship under evolution's ever-changing view.

Some experts believe that the theory of evolution is not very different from other philosophical or religious explanations to origin of life. However, there are gaps in the theory which requires benefit of doubt.

This chapter discusses the origin and evolution of primates.

Figure 2.1. Representation of Evolution of Primates.

Source: Image by Maxpixel

2.2. MEANING OF PRIMATES

The ancestors of primates were those mammals that lived in the tropical regions in the threes. Several characteristics of primates can be attributed to the three-dimensional arboreal environment. Most of these species are atleast partly arboreal. Even the terrestrial ones have tree climbing properties.

Primates are generalist mammals. Their skeletons have developed in a way which allows their limbs to move freely. Hence, they can move around in diverse ways like climb, leap, swing, walk on two or even four legs. They have some unspecialized extremities like long nails and digits which help them perform various functions. Some examples are specialized clawed feet of dogs and cats.

The evolutionary history of the primates cannot be understood from their traditional classification. This is because the classification is based on morphological similarities and not their ancestors. However, some of the traditional taxa still exist hence it is important to understand them.

Traditionally, primates are categorized under two groups:

- Prosimians
- Anthropoids

The former comprised of tarsiers and lorises. Monkeys, apes and humans falls under the latter. Prosimians are smaller and more similar to the first primates. On the other hand, anthropoids were similar and larger in size.

In the present day, primates are categorized based on their evolution. This classification is based on phylogeny, a branch of science, which is concerned with explaining the evolution of a genetically related group of species who have common ancestors.

Figure 2.2. Representation of Primates.

Source: Image by Wikimedia

2.3. ORIGIN OF PRIMATES

Primates are a group of mammals that includes apes, monkeys, lemur and lorises. Even human are primates. Primates have a large brain compared to other mammals. They also have strong arms and legs, long fingers and toes which help them grasp and climb trees. They have their eyes at the front which helps them judge distances with reasonable accuracy.

2.3.1. The First Primates

Primates have evolved recently. Most of the species evolved and became extinct even before the first monkey evolved.

The Earth is about 4.54 billion years old, however, the primates appeared only 50 million years ago.

Towards the end of the Mesozoic Era, creatures similar to primates started evolving around 65.5 million years ago. The Earth then was very different from today. The continents were located in different locations and had different shapes. North America and Europe were connected; however North America was not linked to South America. Asia did not include India, but India was approaching the continent at a rate of 8 inches per annum. Australia was neared to Antarctica than it is today. Most of the continents enjoyed warm tropical or subtropical climates.

The flora and fauna at the end of Mesozoic Era had species which would seem alien today since most of the present-day species had not yet evolved. Gradually, mammals started replacing large reptiles. They started becoming the dominant large land animals.

Though mammals do not lay eggs, there are some exceptions like the archaic egg-layers (monotremes). The platypus and echidna are the descendants of these mammals. Some of the placental mammals existed in those times, which were mainly insectivore ancestors of primates. Majority of the mammal species were small in size, ranging from a mouse to a dog. The cattle and wild beast were not present and developed much later.

2.3.2. Primate-like Mammals

The proto-primates were the first primate-like mammals. Their evolution can be traced in the early Paleocene Epoch. They resembled the tree shrews and squirrels. There is fragmented evidence available which indicates that they were adapted to an arboreal way of life in warm, moist climates.

The evidence suggests that they had relatively good eyesight as well as hands and feet which helps in climbing in trees. Their origins will not be revealed until more fossil data become available. The mammals resembling primates did not contribute to the transformation of terrestrial animal life immediately after the mass extinction that occurred 65 million years ago due to cooler temperatures or competition with the first monkey.

Drastic changes occurred because of the arrival of grazing and browsing mammals. They were equipped with grinding teeth, tough hoofs, and digestive tracts. These helped them process leaves, grass, and other fibrous plant materials. The evolution of these herbivorous mammals paved the way for carnivorous to evolve. They were specialized to eat them. These comprised of the evolutionary lines from which cats, dogs and bears emerged.

Adaptive radiation contributed to the rapid evolution of new species. These happened to increase food getting opportunities and ecological niches. These species mostly comprised of placental mammals.

2.3.3. Early Prosimians

The first true primates emerged somewhere around 55-60 million years ago around the advent of Eocene Epoch. Their fossils have been excavated from Europe, Asia and North America. Their appearance was much different from primates that are found today. They looked like squirrels, but it is believed that they had hands and feet that allowed them to grasp and climb objects and trees. Their eyes were positioned in a way which helped develop stereoscopic vision.

These new primates resembled the present-day prosimians like lorises, lemurs and tarsiers to an extent. The Eocene was the epoch of maximum prosimian adaptive radiation. It comprised of 60 genera split into mostly to families the Adapidae and the Omomyidae.

The former was similar to lorises and lemurs. The later looked like tarsiers and galagos. The diversity of prosimians was atleast four times greater than what is present today. The Eocene prosimians distribution was much wider than it is today. They were spread across Europe, Africa, North America and Asia. During this epoch they spread to the island of Madagascar where they thrived.

The Eocene prosimians were widely distributed because of the no competition from apes and monkeys. These primates, which were more developed, had not evolved by then. Some of the Eocene prosimians were exhibiting evolutionary changes that foreshadow the species that were yet to evolve. The trend was larger eyes and brain, with smaller snouts with each evolution.

Figure 2.3. Representation of Tarsiers.

Source: Image by Wikimedia

There was a hole in the base of the skull and the spinal cord passed through it. This opening was referred to as foramen magnum which meant large hole or opening. The foramen magnum indicates the angle of spinal column which in turn determines whether the body is habitually vertical or horizontal like a monkey. In the Eocene period, it was observed that the foramen magnum started moving to the center from the back. This indicates that they were gradually able to hold their bodies back like the lemurs and galagos. This helped them sit and hop.

These prosimian species started becoming extinct towards the end of Eocene Epoch. Experts link this extinction to the arrival of the first monkey. This occurred during the transition to Oligocene the next geological epoch. This happened 34 million years ago.

2.3.4 Early Monkeys and Apes

The size of several species of primates progressively increased once dinosaurs became extinct. This is because they could flourish on the vast expanses of land and plant food which was more available to them. The Oligocene

Epoch, which was characterized by the evolution of large mammals, was a gap ion primate fossil record, specially the prosimian fossils.

Some of the evidence was found in the Fayum deposits in Western Egypt, a tropical rainforest, around 35 million years ago, but a desert in the present day. Other places where fossil primate bones were found were the Southern Arabian Peninsula, North and West Africa, North and South America and China, Southeast Asia.

It was during the end of Eocene or possible at the beginning of Oligocene era that monkeys evolved. It is highly possible that the evolved from prosimians. The monkeys were the first species of the Anthropoidea. Among the genera of early monkeys, Apidium and Aegyptopithecus, were more commonly known.

Apidium was similar in size to a fat squirrel whereas Aegyptopithecus's size was similar to that of a small dog. Unlike prosimians, they had larger brains, lesser teeth, smaller snouts and forward-looking eyes. The anatomical features of the early monkey indicate that they were active during the day, lived in forests and survived on fruits and seeds.

The New World monkey emerged almost 30 million years ago. The general belief is that they emerged as isolated groups of Old-World monkey that shifted from Africa or North America to South America on large clumps of vegetation and soil. Evidence indicate that the monkeys originated in Africa. It is highly possible that other species of small animals drifted to South America in similar "floating islands" that was created from powerful storms which tore at the lands. This phenomenon is common in tropical regions of the world even in the present day.

Due to lack of adequate fossil records of the Oligocene Epoch prosimians, it is believed that they received competition from the monkeys who replaced them in most environments at that time. This hypothesis is supported by the fact that modern prosimians are mostly found in regions which is not inhabited by monkeys and apes. Moreover, they are generally nocturnal, while more intelligent primates ir diurnal.

The Oligocene epoch was characterized by major geological changes. The regional climate changes changed the course of evolution and affected fossil preservation conditions. During the advent of the Oligocene, Europe and North America were driven away from each other and emerged as distinct continents. Several other changes that took place during the Oligocene had a significant impact on the continental weather patterns. The summer monsoonal rains were redirected to the east. In the Central Asia a huge arid rain shadow region was created and possibly resulted in global climate changes.

The grasslands started expanding during the latter part of the Eocene Epoch, more so in the northern hemisphere. This was a factor which contributed to disappearance of primates from these northern areas. However, climates in most regions were still warmer than today.

Around 14-16 million years ago, which was during the Miocene Epoch, new volcanic mountain chains were created in east Central Asia.

This happened as a result of the movement of f tectonic plates in the Great Rift Valley system. The local weather pattern changed as a result of this. While some areas became arid because of local rain shadow, other became wetter. Besides these, progressive global cooling was being witnessed.

The sea levels were dropping because of less outflow of water from the polar ice caps to the ocean. The coastal lands that were earlier submerged now started emerging. These changes led to the establishment of a land route between Eurasia and Africa Eurasia along the eastern Mediterranean Sea coast. This facilitated the migration of primates and other animals to and from these continents

Dry grasslands and sparse woodlands large part of the South Asian and East African tropical forests. This was the effect of climate changes. Consequently, there were new selective pressures affecting primate evolution.

Primate fossil records have been retrieved from the Miocene Epoch; however, they do not represent all primates in the same proportion. In this epoch, it is believed that apes evolved from monkeys. Fossils of apes are more common from most part of the Miocene, whereas, prosimians and monkeys are rare. There are evidences which indicate that apes at that time occupied some ecological niches that would later be filled by monkeys.

Proconsul was one of the earliest monkeys to ape transitional primates, which lived around 14-21 million years ago in the African forests. The various Miocene primate species included the modern apes and humans. Around 14 million years ago, the group of apes which also includes ancestors of humans was gradually adapting to life on the edges of the expanding savannas in Southern Europe.

It is believed that they were a part of the genus Dryopithecus which resembled the modern Africana apes. Their evolution took place during a brief period of time which experienced a short global heat wave that began 15 million years ago. The polar ice caps started melting due to this heat which caused the sea level to rise between 80-130 feet.

Towards the end of Miocene Epoch, the climate conditions in the northern hemisphere became cooler and hence less hospitable. As a result, the primate species started becoming extinct. A few of them which migrated to South Asia and Africa survived because of the comparatively warm climate.

Around 8-9 million years, in Africa, the progenies of the dryopithecines were divided into two lines. From one the gorillas evolved while the other led to humans, bonobos and chimpanzees. Another divergence occurred, 7 million years ago, which separated the ancestors of modern chimpanzees and bonobos from the early hominins. Hominins were human like primates who were the direct ancestors of humans.

2.4. EARLY HUMAN EVOLUTION

As mentioned, the present-day humans and chimpanzees has its origin from a common ancestor, the hominoid. They diverged around 6 million years back.

2.4.1. Human Evolution

In the primate's order, the Hominidae family comprises of humans and chimpanzees. The fossil records indicate that the chimpanzee and humans had a common hominoid ancestor. This is expected to have taken place 6 million years back. A comparison of the human and chimpanzee DNA also indicates the same.

The evolutionary branch that led to evolution of humans also had many other species. However, humans are the only surviving member of this species. The term hominin or hominid comprises of those species that evolved post the spit occurred in the primate line. As a result, the species that were more close to humans than chimpanzees could be designated.

Though most hominoids were quadrupedal, hominins were the ones who were bipedal. It comprises of the groups that were the ancestors of Australopithecus africanus, Homo habilis, and Homo erectus. It also comprised of ancestral groups like Australopithecus boisei.

Figure 2.4. Representation of Adult Female Homo Erectus.

Source: Image by Flickr

It is difficult to ascertain the true lines of descent in hominins. In the past, very few hominin fossils had been found. In those times scientists believed that considering them in order, from oldest to youngest, would indicate how the early hominis evolved to the present-day humans.

Post that, many new fossils had been discovered. However, the fact that at any one point in time more than one species was alive cannot be dismissed. Hence it is possible that many of the fossils of hominin species found represent species that became extinct and were not the ancestors of the present-day humans. There is another possibility that several new species have been named.

2.4.2. Primitive Hominins

In the recent times, three species of very early hominoids were discussed Sahelanthropus tchadensis, is the oldest of them, which originated some seven million years ago. A specimen of this genus was discovered in Chad. It was basically a skull and was informally termed as "Toumai'. It is a mosaic of primitive and more advanced traits.

Till date, it is not very certain as to how the fossil fits in the picture revealed by molecular data. As mentioned, the species line which evolved into modern chimpanzees and humans divided almost 6 million years ago. It was not perceived at this time that thus species was an ancestor of the modern humans. In other words, it may not have been a hominin.

Another species that was discovered around 5.7 million years ago was the Orrorin tugenensis. This was a recent discovery which was around 2000. Many species of Orrorin have been found. Though it has not been confirmed that Orrorin was a human ancestor, this possibly has also been considered. Some of the traits that Orrorin exhibited was like the present-day humans when compared to australopiths. However, Orrorin is much older.

The third species is the Ardipithecus ramidus which was discovered in the 1990s. They existed around 4.4 million years ago. The scientists who found the first fossil discovered that some of their peers did not support the view that the species was a biped. Hence, they were of the opinion that it is not a hominid.

In the years that followed, various other species of *Ardipithecus* were discovered to be bipedal. This included the new species *Ardipithecus kadabba* which was discovered 5.6 million years ago. It has not been proved yet that this genus was an ancestor of the human. However, since it was a bipedal, it is known that it was a hominin.

2.4.3. Early Hominins

Around 4 million years ago, the hominin Australopithecus evolved. It is believed that they were the ancestors of the genus Homo.

Australopithecus, also called southern ape belongs to the line of hominin. It evolved 4 million years ago in the eastern region of Africa. Around 2 million years ago, it became extinct. Scientists have expressed special interest in the species since it is believed that the Homo genus has evolved from Australopithecus. This development happened 2 million years ago.

Several traits of the Australopithecus resembled the great apes, with less similarities to the present-day man. One example is sexual dimorphism, which was more exaggerated than found in modern humans. The males were significantly larger than males, approximately by 50 percent. This ratio is similar to that seen in the present-day orangutans and gorillas.

On the other hand, the present-day males are also larger than females by 15 to 20 percent. The *Australopithecus* had brains which was proportionally smaller to their bodies when compared to the present-day human. It was more comparable to the great apes.

A common trait between present day human and Australopithecus was bipedalism. It is highly possible though that *Australopithecus* also lived in trees partially.

Hominin footprints have been discovered in Laetoli, Tanzania. These are similar to the footprints of the present-day man. These footprints date back to 3.6 million years ago. These footprints indicate that the during the time of Australopithecus, hominins were walking upright.

Various species of Australopithecus was also called as australopiths. Australopithecus anamensis existed on Earth around 4.2 million years ago. There is another early species called Australopithecus afarensis which existed on Earth somewhere between 3.9 million to 2.9 million years ago.

This species bore evidence to a trend in human evolution. The dentition and jaw were gradually reducing in size. A. afarensis had smaller canines and molars in comparison to apes. However, when compared to that of modern humans, it was larger. Its brain size was 380-450 cubic centimeters which is similar to the size of the present-day chimpanzee brain. Its jaw was prognathic, which is larger when compared to that of the present-day humans. Around the 1970s, the fossil of an adult female A. afarensis was discovered. The fossil was found in a remote region of Ethiopia.

This fossil was termed Lucy. This is important because it was one of the most conclusive evidence of australopiths fossil discovered. Almost 40 percent of the skeleton was discovered.

Australopithecus africanus existed on Earth somewhere between 2 to 3 million years ago. Its built was slender, but arms were strong. One of the reasons behind this may have been the significant time that they spent on trees. Its brains were bigger in size when compared to that of A. afarensis. It was estimated at 500 cubic centimeters, the human brains are more than three times this size.

Australopithecus bahrelghazali and Australopithecus garhi are to other species which are now a part of the roster of australopiths.

The majority of the australopiths had a more lean and gracile build. Their teeth were more suited for chewing soft food. There were some australopiths which had a more robust build. They were found almost 2.5 million years ago. These hominids were larger, and their teeth were fit for grinding. Their molars indicate that they mostly consumed fibrous and coarse vegetarian food. It showed heavy tear. Whereas the more gracile australopiths had atleast some carnivorous food in their diet. These include *Australopithecus robustus* of South Africa, and *Australopithecus boisei* of East Africa and *Australopithecus aethiopicus*.

These species disappeared from Earth almost 1 million years ago. They are not considered as ancestors of present-day humans. Instead they are part of an evolutionary branch on the hominin tree which has no descendants today.

1. Genus Homo

The genus Homo comprised of modern humans and also its extinct humankind. They first appeared on Earth 2.3 million years ago.

The fossils of the species called Homo habilis was the oldest example in the genus homo for many years. In the year 2010, Homo gautengensis, a new species was proposed. It was considered that it could be older, although this fact has not been established. Several traits of H habilis was comparable to the modern man, which was more than what Australopithecus africanus had.

The jaw of H. habilis was less prognathic, i.e. its jaws were not that projected when compared to australopiths. It also had larger brain which was around 600-750 cubic centimeters. This H habilis however had a few features that the older hominin species had, one of which was long arms. The term H habilis is interpreted as handy man. This term was coined because their remains had some stone tools.

H. erectus emerged on Earth approximately 1.8 million years back. Scientists believe that the first appeared in East Africa. They were the first hominin species that drifted from Africa. It is believed that they spread to China, India, Java and Europe as the fossils have been found in these places. They were earlier called as 'Perkin Man" or "Java Man'.

This H-erectus had more traits in common to the present-day man than the H. habilis had. The H. erectus were larger when compared to the earlier hominin. Fossils indicate that they could have been 1.85 meters tall, weighed around 65 kilograms. Their sizes are comparable to the present-day man. Its degree of sexual dimorphism was less when compared to earlier species.

The males where 20 to 30 percent larger when compared to females. Similar size difference is observed in the present-day human.

The brains of *H. erectus* was larger when compared to some earlier species. It was around 775 – 1,100 cubic centimeters. This is comparable to the present-day human brains which is around 1,130 – 1,260 cubic centimeters. The nose of the H. erectus had downward facing nostrils. The same trait is observed in the present-day human. Whereas the other primates have a forward-facing nostril.

The downward-facing nostrils would have evolved to allow the cold air to become warmer before it gets into the lungs. This could have evolved as a means to adapt to the cooler temperature. Various artifacts were discovered along with the fossils of *H. erectus*. This indicates that H. erectus were the first of hominin that used fire. They also used to hunt and have a base. H. erectus existed on Earth till around 50,000 years ago.

2. Homo sapiens

The H. erectus gave birth to many species, around 50,000 years ago. They are referred to as Homo sapiens. The brains size of the H. sapiens was larger compare to H. erectus and measured around 1,200-1,400 cubic centimeters. However, some of their traits differed from the present-day humans. They had a thick skull, a receding chin, a prominent brow ridge. A few of the species were around even 30,000 to 10,000 years ago. It is believed that for some time, they could have co-existed with the present-day humans.

There is little agreement over ho the anatomically modern homo sapiens first emerged. As mentioned, the H. erectus drifted from Africa into Europe and Asia when the first major wave of migration occurred around 1.5 million years ago. The multiregional hypothesis proposes that humans first emerged in the beginning of the Pleistocene era almost two million years back. The subsequent human evolution that has occurred has been confined to a continuous human species.

This species includes archaic human forms like Homo erectus and Neanderthals and various modern forms. They spread to other parts of the world and evolved into the modern human sapiens. This hypothesis proposes that humans have evolved through a combination of adaptation that are found in many parts of the world. This is possible through gene flow between those regions. Proponents of multiregional origin supported their hypothesis by using genomic data and fossil and continuity of archaeological cultures as evidence.

The primary alternative hypothesis explains how the present-day humans originated in Africa. This theory holds that the present-day humans evolved in Africa around 100,000 – 200,000 years ago. They drifted away from Africa around 50,000 – 60,000 years ago and replaced the archaic human forms with limited interbreeding. It is believed that this interbreeding has happened at least once with Denis vans and once with Neanderthals.

2.5. ADAPTATION

Living beings acquire adaptation through evolution. They usually carry some type of advantage that enables a species to transfer its genetic material to the next generation. Adaptation generally takes one of the three forms namely, structural, physiological a behavioral.

2.5.1. Structural Adaptations

A structural adaptation is related to the physical aspect of an organism. The physical change often occurs as a result of any modification in the organism's physical environment. For example, if the ecosystem suddenly develops forests, the animals living there will develop climbing claws and suction pads. This would grant them a distinct advantage over the animals which did not evolve. Similarly, development of wings helps organisms fly, string legs help them jump and fins help them swim.

2.5.2. Behavioral Adaptations

Behavioral adaptation involves a change that influences the way a species acts and reacts to its surrounding. This type of adaptation is often a result of change in the surrounding environment or because of any actions of another species. For example, some predatory animals might form packs to hunt. This gives them an advantage over predators that hunt alone. The other behavioral adaptations include modification in social behavior, communication pattern, dietary habits and reproduction.

2.5.3. Physiological Adaptation

Similar to structural adaptation, physiological adaptation involves a physical modification in a species. The physiological adaptation is not always evident in the appearance of the organism. This type of adaptation is caused either due to a change in the environment or because of actions of other species.

For example, a species that lives near water may become more acidic and gradually adapt by slowly changing its own body chemistry. Some of other adaptive traits include greater intelligence and heightened senses.

2.6. STRUCTURAL ADAPTATION

The book "On the Origin of Species" by Charles Darwin, he did not discuss the evolution of humans. He thought it would be a controversial topic as he did not have enough data to support his arguments. A decade later, Darwin wrote a book "The Descent of Man" which as he suspected started the long-lasting deliberations and made evolution a controversial subject.

In this book, Darwin, researched the special adaptive features which was seen in various types of primates like lemurs, apes, gorillas and monkeys. These adaptations were structurally similar to what humans have. During that time, the technology was not advanced much which led to criticization of the theory by many religious leaders. In the last century, more fossils and DNA evidence were uncovered which supports Darwin's ideas that he proposed regarding adaptation in primates.

2.6.1. Opposable Digits

The primates possess flexible five digits in their hands and feet. The primitive primates used them to grasp branches of trees where they dwelled. One of these five digits is not aligned with the other digits, rather it sticks out of the hand or foot. This digit is known as the opposable thumb in case of hand and opposable big toe in case of foot. The earliest primates made use of the opposable digit solely for grasping branches as they swung from one to another. Gradually, they started using it for other purposes like grasping tools and weapons.

2.6.2. Fingernails

Any animal which has individual digits in their limbs has claws attached to the end. This is used for scratching, digging or for protecting themselves. Primates have nails which are flatter and are covered with keratin. These fingernails and toenails also shield the flesh at the tip of the toes and fingers which are delicate and fleshy. These areas help them sense when they come in contact with something as the fingertips are sensitive. This primarily helps in climbing trees.

2.6.3. Ball and Socket Joints

All primates have joints in shoulder and hip. This is referred to as ball and socket joints. The name itself suggests that it comprises of a bone in the pair. While one has a rounded end similar to all a ball. The other one has a caved in space where the ball fits in, similar to a socket. The primates can climb the treetops easily due to this adaptive feature. The socket allows a 360-degree rotation of the limb. This adaptation facilitates the primates to climb with less efforts and reach the top of the tree easily in search of food.

2.6.4. Eye Placement

The eyes of the primates are on the front of their heads. Most animals have their eyes located on the side of their heads which aid in better peripheral vision. Some have it on their top so that they can see even when they are submerged in water. The primates have an advantage since both their eyes are located on the front. Both the eyes can grab the visual information from both the eyes simultaneously and help the brain put together a stereoscopic or 3-D image. Their vision helps them judge the distance and perceive depth. This is especially useful for climbing trees or jumping to higher branches without falling. This is particularly critical because misjudgment about distance or height can even lead to their death.

Figure 2.5. Eyes of the Primates are Set in the Front which Helps in Grabbing Visual Information Simultaneously.

Source: Image by Nicepic

2.6.5. Large Brain Size

A stereoscopic vision could have resulted in a relatively large sized brain. The primates have to process all extrasensory information. Hence the brains have to be larger so that it can perform all the necessary work simultaneously. A large brain not only helps in surviving, it also makes them more intelligent and help them develop social skills. Primates prefer being social. They live in communities and work together which makes their work easier. Therefore, primates live for a longer period of time, develop maturity much later and nurture their young ones.

2.6.6 Adaptations that Facilitate Locomotion

1. Primate Grasping

Grasping is perhaps the most remarkable adaptation observed in primate limbs. The primates possess the ability to hold onto small curved surfaces. This is mainly useful in holding tiny branches. This facilitates the primates to walk around and discover even the smallest detail of the arboreal canopy. Some distinguishing features of primates are nails while most animals have claws. It possesses many large intrinsic and extrinsic muscles. This is helpful in digital flexion and grasping. Hand and foot rotations are possible because of their mobile joint surfaces. Primates also possess wide fingers and toes which have broad palms or soles. This broad surface is contributing stability during arboreal locomotion when in contact with arboreal substrates. All these features are significant for grasping.

The primary function of Palmigrade hands of primates help inactive grasping and climbing structures. Besides they also reflect exhibit many postural types. This includes digitigrade, palmigrade, fist-walking, and knuckle-walking, suspensory hand positions. African apes and orangutans fold their long fingers underneath when they move quadrupedally in terrestrial areas. This is possible due to fist-walking and knuckle-walking hands.

Primates hands also help them to gather and consume food. Another aspect of primate hand adaptation allows grasping, reaching, and pulling things close to their mouths. This feeding adaptation has figured prominently in the visual predation hypothesis of Cartmill (1972). The secondary evolution of claw-like nails among a variety of different primates has been linked to both the ecological use of large-diameter trunks and to exudate feeding. This is a substrate situation that is much beyond the grasping span of hands and feet (Cartmill, 1979).

2. Climbing

Similar to grasping, climbing is ancient arboreal adaptation noticed in primates. It is observed that primates move up vertical substrates. There are certain quantitative studies which have demonstrated that climbing is not particularly frequent movement in any locomotor profile. The most dominant ones are Quadrupedalism, leaping or brachiation are more dominant profiles.

Primates clearly need to climb upward in trees but relative to all other movement types, vertical climbing is not frequent relative to the traveling modes of locomotion. Apes is one of the climbing primates which climb vertically when arboreal. However, the forelimb lengthening, and their highly modified upper body has probably evolved from their ancestors who used to swing their arms.

Some experts classify the locomotor categories into two types: vertical climbing and clambering. While the former refers to going up or down a vertical support, the later refers to oblique movement through interlinked smaller branches. They differ in the support in which primate climb up and the angle in which they ascend. Vertical climbing is not as common as clambering.

An unusual movement pattern in climbing is bridging. In bridging, a primate covers the gap and pulls itself to a branch on the other end. Orangutans, atelines and lorises make use of this method to bridge the distance in the canopy.

3. Quadrupedalism

Quadrupedalism falls under two categories arboreal and terrestrial. The former is more common. The fore and hind limbs of the primates displaying quadrupedalism are of the same length. They generally bend their elbows and knees to lower their centers of gravity towards the branch. Primates use diagonal couplets because of which primates can have only one limb off the substrate at a given point. Arboreal quadrupedal primate supinates their hands and feet toward the curve of the support. This is also known as palmigrade support. One the other hand, the hand and feet terrestrial quadrupedal primates are same. However, these primates, Old World monkeys in particular do not have great joint mobility in their entire limbs. Moreover, they utilize more pronated hand positions.

There are four categories of terrestrial quadrupedalism namely, knuckle-walking, digitigrades and fist-walking. The African apes make of use of terrestrial quadrupedalism by folding fingers at the first joint. Knuckle-walking is the term used for this type of movement. Their arms are longer than their legs and their back is angled at 45 degrees.

The movement observed in orangutans is a fist-walking hand posture. They generally have highly supinated foot positions. It is observed that the hand position of terrestrial Old Monkeys is often digitigrades.

4. Leaping

The leaping primates fall under two categories. The strepsirrhine primates and tarsiers exhibit strong upward parabolic leaps. The anthropoids on the other hand tend to leap outward at the same level of their body and fall downward. Both these categories leap habitually. The only limiting factor tends to be the size. For examples, primates weighing more than 10 kg seldom leap. The primates have long legs which help them leap at a great height and distance. The leaps have less relative muscle force.

The leapers have long femora. The distinguishing factor between an occasional and regular leaper is the tall antero-posterior height and high lateral patellar rim. These animals have large muscles of quadriceps, which produces the required leg force which helps them extend and makes their movement swift and efficient.

Tarsiers have extreme hind limbs that have tibiofibular fusion. Their foot bones are also long. The galagos extend their foot length by elongating feet bones. This is helps them keep their legs longer and have efficient level mechanism.

Experts point out that primates elongate their calcaneus and navicular bones higher up in the foot instead of their more distal metatarsals. This is a morphological compromise between the efforts required for grasping and leaping. Long lever arms generate better gear ratio which helps them leap to great distances. The feet of arboreal primates must enable them to grasp. The galagos and tarsiers extend their calcaneus and the navicular in order to remain mobile toward the digits at the same time extend their foot length.

5. Brachiation

An unusual primate movement which is observed in case of primates which have dominant hind limbs is arm hanging and arm swinging. This feature has evolved atleast, as noticed in apes and spider monkeys. The torso of all apes existing today, as well as humans differs greatly from those of other primates. These upper body features are associated to arm suspension and brachiation.

It is seen, the thorax of apes is broad and also flat in an antero-posterior manner. This is an orthograde or an erect back relative to the common pronograde or horizontal backs of primates. The shoulders of these primates are pushed out to the sides of the body. The scapulae lie on the back wall and not along the sides of the rib cage unlike what is seen in quadrupedal primates. This is a novel scapular position and pushes the shoulder joints away from the midline and to the sides of the body. This contributes towards improving the rotational mobility of the arm.

The clavicle's length is enough which helps it attain a new shoulder position. This is because the forelimb is of considerable length. Consequently, stride length or arm-swing, as the case may be, is also long. The primates cannot twist their enlarged humeral head joint because of the positioning of their shoulder. As a result, their elbows face forward.

2.7. BEHAVIORAL ADAPTATION

The primates do not generally modify the environment they live in, instead they use it without changing it. The great apes have sleeping nets which are roofless, poor and created on a temporary basis. Monkeys do not build nests instead they sleep on trees.

Humans are the only primate who store food. The adult male chimpanzee and adult bonobos both male and female form groups with each other and hunt for food. They do not store food; however, they share food to socialize with others in their community. It has been observed that a chimpanzee carries carcass almost over for a mile to reach the threes safely and eat then it.

Most of the non-human primates do not manufacture or use tools. Some exceptions are common chimpanzees, gorillas, orangutans, bonobos and capuchin monkeys. Some of them acquire food and water using simple tools. For example, chimpanzees pluck leaves from twigs to make probes which they use to eat ants and termites.

They also use modified sticks to eat home directly from beehives. They also use twigs as toothpicks. Chimpanzees use crinkled leaves as sponges and obtain drinking water from hollows in trees. They also use stones and tree branches to crack open nuts. They are also sometimes hurled at other animals to intimidate them.

These simple tools are generally taught to baby chimpanzees by adults and not inherited genetically by the children. Hence, different communities use various tools and as more species are carefully observed, other examples of such tools would be discovered.

A major advantage that that primates possess is the close association of the old and young throughout their life. They live in societies where such close association is possible. This allows the young to learn survival strategies from older chimpanzees who are experienced and possess more knowledge. This equips the primates with the required skills by the time they grow. They interact efficiently with each other and the environment.

The primates may not have developed instinctive survival skills, but their social skills are effective. When they are acting in groups, they can tackle with their immediate predators. Groups of primates can also discover and control food sources better. However, there are exceptions where individuals lead solitary lives.

Most of the primate have a vegetarian diet. Non-human primate groups look for plant foods in dispersed patterns within their own territory. Hence there is a high possibility that none of the food sources go unnoticed. Foraging in tightly spaced groups is effective in cases where food is found in concentrated clusters. Chimpanzees are the only primates that use calls to inform others when they discover food. This message is relayed to the entire group only if there is sufficient food for everyone.

Non-human primates do not share food easily except with infants. Feeding monkeys do not allow others within a certain distance. When a monkey sees another monkey of the community eating food, they rush towards them to grab some. Even if a monkey does not want to, it communicates the presence of desirable food through its actions. This way the information about food sources get shared, especially in case of macaques and baboons.

Chimpanzee are an exception to the above. The males form small groups and hunt for meat. When the group kills a target animal, the primary hunter claims monopoly over the meat. When there is enough meat for everyone, the primary hunter shares it with others and even children and females who were not a part of the hunt. Chimpanzees o not generally share vegetable foods which they can obtain more easily except by mothers and their children. It is the mothers who teach their children what food is edible in the environment. The mothers also teach them important skills like nut cracking etc.

Among chimpanzees and baboons, the size and structure of the foraging group modifies itself based on the size of food supply. Monkeys which are arboreal are not equally flexible. The whole community comprises of a single foraging group. This can be attributed to the fact that their forest territories have more rich vegetable foods and are smaller in size.

2.7.1. Intelligence in Primates

In comparison to other animals, primates have larger brain to body ratio. They also exhibit higher intelligence. This is specifically true as far as great apes and humans are concerned. Chimpanzees can recognize their reflection in the mirrors. This kind of self-awareness is not common among animals. Even monkeys are not this intelligent. It was earlier believed that only humans possessed the intelligence required to recognize their own reflection.

Humans share some of their self-awareness with atleast few of the great apes. This is not surprising because there is a very short evolutionary distance between humans and the apes. Humans had a common ancestor with the chimpanzee 6-7 million years back.

In the past, biological anthropologists were of the opinion that the high intelligence that was found in humans and great apes evolved because of the process of natural selection which was more incline towards tool making. It is more evident now that tool use is relatively modest compared to humans and even some insects. Examples are termite mounds and beehives which are more sophisticated when compared to what apes create.

Many experts believe that apes have large ape brain because it helped them deal with complex social interaction. The other purpose besides this was to aid in tool making and manipulating objects in their surroundings. The connection between intelligence and social skills can be seen operating in chimpanzee societies.

In order for an individual to succeed, he or she has to possess the skill to learn from others in their environment, adopt new behavior, and understand the status and complex emotions of others in the group. This knowledge helps them gain an advantage over others. Chimpanzees can acquire the skills to influence and also modify the emotions of other species in their communities.

The male adults, of high rank in particular acquire the skill to bluff, intrigue and deceive others. They are calculative animals who engage in politics. They spend a good part of their time creating bonds with other males and maintaining them. In order to succeed, they have to have considerable influence on their social worlds.

It has been witnesses that the non-primates display behavior pattern which are not based on instinct. This is particularly in case of great apes. They depend on their acquired skills and knowledge which help them survive. Various anthropologists did not refer to this as culture. Instead they used this term for more complex human knowledge, skills and traditions.

If culture is perceived as learned behavior patterns, then it can be accepted that a few of the non-human primate communities have cultural knowledge that they transfer to their children informally, similar to the way humans do. The children acquire the survival skills mostly from their mothers or other adult females. This is observed in all primate species including humans.

Some experts point out that, cultural knowledge of non-human primates is still very basic. Hence it can be termed as protoculture. In other words, they have the beginnings of culture.

2.8. PHYSIOLOGICAL ADAPTION

One of the physiological adaptations that primates exhibit is dietary caloric restriction. It helps in effectively and reproducibly slow down age. This helps in maintaining vitality and health in mammals. This paradigm has been established over 6 decades back. However, how it functions and is applicable to humans is not known. Various studies were conducted which point out that the benefits of anti-aging and anti-disease is also applicable to humans.

Some primates live in high altitudes. One example is the golden monkey, which is a special species in China. It possesses the highest altitude habitat. Primates like this monkey have shown physiological adaptations which help them adapt to such a high altitude and withstand extreme conditions. The reproductive, respiratory and blood systems are evolved which help them adapt to these conditions.

Another phenomenon observed in primates is hibernation. During hibernation, the body temperatures of the primates fluctuates between 10°C and 30°C. It closely tracks the diurnal fluctuations of the ambient temperatures passively.

REFERENCES

1. https://www.allaboutscience.org/what-is-evolution-faq.htm [Accessed 27 Aug. 2019].
2. https://www.ck12.org/book/CK-12-Human-Biology/section/7.2/ [Accessed 27 Aug. 2019].
3. https://courses.lumenlearning.com/boundless-biology/chapter/the-evolution-of-primates/ [Accessed 27 Aug. 2019].
4. Roth, G., Ingram, D. and Lane, M. (2006). Caloric Restriction in Primates and Relevance to Humans. Annals of the New York Academy of Sciences, 928(1), pp.305-315.
5. https://sciencing.com/three-types-environmental-adaptations-8572825.html [Accessed 27 Aug. 2019].
6. https://www2.palomar.edu/anthro/earlyprimates/early_2.htm [Accessed 27 Aug. 2019].
7. https://www2.palomar.edu/anthro/behavior/behave_3.htm [Accessed 27 Aug. 2019].
8. https://www.ncbi.nlm.nih.gov/pubmed/15146651 [Accessed 27 Aug. 2019].

Chapter 3

Taxonomy and Classification of Primates

"Taxonomy is described sometimes as a science and sometimes as an art, but really it's a battleground."

— Bill Bryson

CONTENTS

This chapter named as 'Taxonomy and classification of Primates' provides insights about the classification done in the primates and the taxonomy in primates to the readers. In this chapter, in the first part, the history of classification has been mentioned. This chapter provides the proper classification of primates along with a table to make the reader understand about the order and sub – order, family and many more. Each sub order and family of species have been proper described in this chapter.

3.1. INTRODUCTION

> Taxonomy can be defined as studying the principles and practice of classification, and is widely used as a synonym for systematics, although the latter one, in practicality, broader, covering all the aspects of relationships among organisms.

The term taxonomy is derived from the word taxon (plural taxa).

A taxon can be defined as a classificatory group, and it consists of hierarchy: genus (plural genera), species, order, family, class, kingdom and phylum (plural phyla). The kingdom Animalia (the animals) is categorized into different types of phyla, which comprises of Chordata (more or less, the backboned animals).

Each phylum is further divided into different classes; one of the commonly known or recognized class of the Chordata is the Mammalia (the mammals). Each class is dived into different orders or patterns; Order Primates is one of the orders of the Mammalia. Each order comprises of different families, each family then split into different genera, each genus into different species. (It may also be the case that there is one and only known species in a particular genus, only one identified genus in a specific family, and so on.)

The hierarchy is thus considered as a nested one. These seven ranks of the taxonomic hierarchy are known as the obligatory ranks: every species of animal is categorized into a genus, each genus then into a family, and so on up.

In case the finer degrees of classification are required, the obligatory ranks can be segmented into different parts: suborders, then infraorders, then parvorders, which is put in below the rank of order, subfamilies can be divided into different large families which comprises of many genera, and the rank "tribe" can be put below subfamily, and other subordinate ranks can be inserted.

It is important to note that the different groups of families can be combined and segmented into a superfamily. Further, these "sub-" and other subordinate categories should only be inserted when they are needed; but it is obligatory to list the seven ranks mentioned earlier.

The International Code of Zoological Nomenclature is responsible for regulating and managing the naming of families, species and genera. The rules, as it is framed in the Code, bring order where there has been significant chaos and complications in the past, and have stabilized nomenclature. The name of a genus is a Latin or Latinized singular word and is always found in italics (just as the name of a species), and always started with a capital letter.

In addition, the name of a species is also Latin or Latinized; it may be a noun in opposition, an adjective, or a genitive (such as orschmidti (of Schmidt), katanganus (of Katanga), and always started with a small letter and, also combining with the generic name, forms a binomial which distinctively entitles the species in question (Eulemur mongozn and Macaca fascicularis).

It is usually seen that sometimes what is in effect the same taxon turns out to have been explained or signified more than once, under two or more than two names. These names are known by synonyms, and the rules or guidelines of nomenclature allows taxonomists to take decision with respect to the synonym that is applicable in the case or right to be used (in principle, the one described earliest).

The name of a family is resulted from the name of one of its included genera, and ends with -idae (such as Cercopithecidae, the family that entitles the Old-World monkeys, the name being obtained from Cercopithecus, one of its included genera). The names of subfamilies, if they are required, end in -inae; if still further subdivision is required, tribes are recognized within subfamilies, and they end in -ini.

It is also important to note that the names of super families usually end in -oidea. There are no predefined endings for orders (or suborders) or of some higher categories. It should be kept in mind that the codified rules of nomenclature must be followed by all biologists, in a similar manner in which they are designed with the objective to provide standardized names for potentially valid taxa.

In case a taxonomist finds certain evidence that two genera, formerly thought or predicted to be different, are actually the similar one (i.e., they are synonymous), then the rules or guidelines of nomenclature explain the taxonomist about to use which name for the newly combined genus.

In a similar way, if a taxonomist finds that a single genus ought to be split into two, then again, the rules of nomenclature will help in explaining the taxonomist, out of two, which should bear the name of the old, pre-split genus, and what the name of the other genus should be.

As a general rule, "priority" is the watchword: the earliest name (recognized by the name senior synonym) should be used, although sometimes for one reason or for another the International Commission on Zoological Nomenclature rules that only one of the junior synonyms should be used.

The International Code of Zoological Nomenclature is now in its fourth edition (which came into effect at the starting of the year 2000). There is the need to incorporate fossils into this scheme.

Whereas, it is generally seen that some fossil taxa (the Proconsulidae, for example) have a significant time depth and have their own set of defined derived features, and can be segmented in alongside with the families thought to be their closest relatives in the living fauna, others (such as Morotopithecus) were generally of very less time duration and, having no traceable derived features, could even be common ancestors to two or more than two living (or long-lasting) taxa.

It is usually recommended that these potentially ancestral and short-lived taxa should be left unranked and given the status of "Plesion." In case there is any evidence to explain that some ranking "universally accepted" is wrong, then there is a need to change the classification—as long as this new evidence is good as it is required, and it does not look as if any future discoveries will alter things.

Taxonomy is a specialized field of endeavor, and it is also perfectly viable for non-specialists to practice the science, training and good knowledge in the field is required. In particular, it is generally seen that the rules of nomenclature are poorly understandable outside the field, as is species theory.

3.2. HISTORY OF CLASSIFICATION

An approach was developed by Aristotle to nature for the purpose of classification of living beings in 4th century B.C. in this, the categories were based on the visible similarities and one member of each category was considered as primate. This word primate has been derived from the Latin word Primus which means the first or best of the Group.

With reference to the discoveries that were going on around the globe and in the inner working of the organisms, Swedish naturalist Carl Linnaeus developed the Sistema Nature which is a system of nature in the 18th century so as to classify all the living things.

Figure 3.1. Swedish Naturalist Carl Linnaeus.

Source: Image by Flickr

This collection reflected a new understanding about the place of man among the animals. Linnaeus had the similarity among humans, monkeys and apes which further classified them together as primates. The basic biological unit of the classification is the species. The largest normally interbreeding population is referred to as a species. The name of the species is made up of two words that is a capitalized generic name and a trivial name.

This system of classification is called as Binomial Nomenclature. There is Genera which usually include the species which are very much similar, but the difference is in the body size or some relatively minor features of body proportions, behavior, coat color etc. They are clustered into families. Furthermore, families are divided into sub – families or clustered into super families.

The family group name nomina also follows a definite pattern. Family name ends in suffix – 'idea' for example, Hominidae however, the super families end in – 'oidea' for example, Hominoidea. The modern taxonomy or the science of the classification is based on more than body structure function and growth.

3.3. CLASSIFICATION OF PRIMATE

The living primates are divided into two sub – orders. Recently, the sub – divisions have been the lower primates or Prosimian and the high primates or Haplorrhines.

3.3.1. Strepsirrhini / Prosimians

Strepsirrhini or Prosimians are the most primitive of the living primates, which also include the monkeys and apes. The meaning of the name prosimian is pre – monkey. The living prosimians are all placed in the suborder Prosimians which further include four families of lemurs (the Lemuridae, the Cheirogaleidae, the Indriidae, and the Daubentoniidae), the bush babies, lorises and pottos (family Lorisidae), and the tarsiers (family Tarsiidae). There are some other authorities which also include the tree shrews, though others separate the tree shrews into their own order.

Figure 3.2. Ring Tailed Lemur Prosimians.

Source: Image by Pixabay

It can be said that Prosimians are primarily tree – dwellers. They have a longer snout in comparison to the monkeys and apes and also the prosimian snout usually has a moist nose which indicates that they have a well – developed sense of smell. Much larger portion of the brain of prosimians is dedicated to the sense of smell in comparison to the sense of vision.

Prosimians actively scent – mark their territories in order to warn other animals of their occupancy. The scent – marks are made with a fluid which is strong – smelling which are produced by special glands or with urine or feces. The eyes of the prosimians are large and are adapted for night vision. They have a tapetal layer in the retina of the eye which reflects and reuses light. Prosimian eyes are not as well positioned for the stereoscopic vision similar to the eyes of other primates.

Similar to all the primates, prosimians have hands and feet which are capable of grasping tree limbs. The second toe of the hind foot of prosimians has a long claw. They use this claw for grooming. The other toes on both the hands as well as the feet have flattened nails and no curved claws. Lemurs walk along the branches on all the four legs and they have longer hind legs than the front ones.

Tarsiers are adapted to leap in between vertical tree trunks and then clinging to them and also, they have short legs whose bones are fused together for the purpose of strength.

Prosimians have inflexible faces in comparison to those of monkeys and apes. There are most of the prosimians which have 36 teeth, but west simians generally have 32 teeth. The lower front teeth of prosimians lie horizontally and protrude thus, forming a grooming structure which is called as dental comb. This dental comb is further used in order to comb fur and scrape while nourishing the gum from trees. After this, it is cleaned with a hard structure which is located beneath the tongue. Prosimians spend much less time in infancy than the simians do, perhaps only about 15 % of their lifespan with regards to 25 to 30 % for monkeys as well as apes.

The early primates were distributed throughout most of the portion of the world. Now – a – days, the majority of the living prosimians, are the ones which are collectively called as lemur and they live only on the large island of Madagascar, Africa. After the human beings arrived on the Madagascar that is about 1,500 years ago, there are at least 14 species of lemurs have become extinct.

There are smallest living Madagascar prosimians which are the mouse lemurs in genus Microcebus, however, the largest lemur is the indri (Indri

indri). The other prosimians which do not live in Madagascar, and they fall into the groups such as the lorises, pottos and galagos or bush babies of Africa, India and Southeast Asia and the tarsier of Southeast Asia.

Lemurs

The lemurs belong to a group which is called as prosimian primates which can be defined as all primates which are neither monkeys nor they are apes. However, there are a number of species of lemur and there are very few individuals. Lemurs are considered as the most endangered group of animals on the planet. These kinds of primates are the only found in one small area of the earth and there are many species which have small and decreasing numbers.

Size

There are a large number of species which are around 105 species of lemur and they naturally come in a wide range of sizes. The largest lemur species is the indri. These species grow up to 24 to 35 inches that is 60 to 90 centimeters long and they weigh 15.5 to 22 lbs that is 7 to 10 kilograms. Once, there were lemurs which were a big like gorillas. But these are now extinct.

Habitat

Lemurs live in one place only in this world, Madagascar and the places nearby Comoro Islands which are located off the coast of Mozambique in Africa. They occupy a number of various habitats like the dry deciduous forests, spiny forests, rain forests, wetlands and mountains.

Figure 3.3. Sifaka lemurs.

Source: Image by Wikimedia

Habits

Lemurs are considered as a very social creatures and they live in groups which are known as troops. The ring – tail lemur's troop is led by a dominant female and thus, can include 6 to 30 animals and this fact is according to the National Geographic.

When the lemurs are awake and active, this depends on their size. Typically, small lemurs are nocturnal which means that they are active at night. Whereas, the larger lemurs are considered as diurnal which means that they are active during the day time.

There are some nocturnal lemurs which live alone but they are still social with other lemurs. They will set off on their own during the waking hours, but they will sleep with the groups of other lemurs. A large number of lemurs spend their awake time on trees. In case, if they are not eating then they like to groom each other or have a sunbathe.

Diet

There are some lemurs which are herbivores which means they do not eat meat. They love fruits and they can also eat flowers, tree bark, leaves and sap. Some other lemurs are omnivores that means they eat a variety of foods which include fruits, nectar, flowers and leaves and also the insects, spiders and small vertebrates. This fact is provided by the World Animal Foundation.

Lemuroidea

The Lemuroidea is considered as a super family which inhabits Madagascar Isle. This super – family is represented by Lemurs. They are arboreal creatures having nocturnal habits. There seems to be a link in between the sleeping habits or the walking activity, size of the eyes, body size and social habits in case of smaller lemurs such as lepilemur. They have the biggest eyes. They generally live in small social units or in family groups.

Figure 3.4. Lepilemur Hubbardorum.

Source: Image by Wikimedia

On the other hand, the larger lemurs such as indri are frequently quite active by day or in twilight. They have small eyes and often tend to live as members of a community of their own species. By this, it can be said that the increase in the size of the body can be a deciding factor in order to abandon their nocturnal or crepuscular habits in the early primates.

Lorisoidea

The Lorisoidea is found in Africa and Asia and they both differs from other Prosomians in a number of various ways. Their size is smaller, but they have larger ears and eyes. Along with this, they are more or less tailless, moving slowly with the rudimentary or wholly reduced index – finger. The small second toe has a long cleaning claw. The Lorisoidea is a super family which consists of various genera.

There are some characteristics features which are mentioned as:

- The Potto or Perodicticus is the size of a small cat.
- It is further categorized by a unique specialization.
- The spines of the last cervical and the first two thoracic vertebrae shows through the skin which is capped by the horny spines which contains a formidable defense weapon in oppose of attackers.

The size of the Bush Babies that is Gagidae is like a rat which is characterized by the presence of long bushy tails, short muzzles, large mobile ears and much elongated ankle bones that make them remarkable jumpers in all the primates. The dwarf bush baby that is Galagoides are the ones which nestles easily on the palm of a man's hand and they can leap a dozen or more feet in any direction including the upward direction.

Figure 3.5. Galago Bush Baby.

Source: Image by Pixabay

The head can be turned around so that the nose is in line with the vertebral column. The digits of the hands as well as feet are slender and long. The pollex and hallux are well – developed and so is opposed to the other digits. The pelvic limbs are much longer in comparison to the pectoral limbs. The lorises are not able to leap or climb by hand over hand motion. The galagoes do progress by rapid saltation.

3.3.2. Haplorrhines

This particularly refers to the suborder Haplorrhine of the order Primates. The suborder Haplorrhini is a suborder of primates which mainly comprises of the following families: Callitrichidae (tamarins and marmosets), Tarsiidae (tarsiers), Aotidae (owl monkeys or night), Cebidae (squirrel monkeys and capuchins), Pitheciidae (sakis, titis, and uakaris), Cercopithecidae (Old Word monkeys), Atelidae (spider, howler, and woolly monkeys), Hylobatidae (lesser apes), and Hominidae (humans and great apes).

Figure 3.6. Tarsier.

Source: Image by Flickr

It is usually seen that the haplorrhines are mainly primarily known by their lack of rhinarium. The rhinarium is the surface of the skin that covers the external openings of the nostrils. It is also available in another subgroup of primates, the strepsirrhines. The haplorrhines are mainly known as dry-nosed primates.

They have a dry area between the nostrils and the upper lip. Another unique feature of haplorrhines is the lack of a functional terminal enzyme that primarily perform the function of producing vitamin C. This enzyme is in existence and functional in strepsirrhines. Thus, the haplorrhines must accumulate their vitamin C from their diet. It is also observed that the brain-to-body ratio of haplorrhines is higher as compared to strepsirrhines.

1. Anthropoidea

As many of the early prosimians gradually became extinct at the end of Eocene epoch, new breed of primates came into existence in tropical forest environments. The divergent evolution of the higher primates is closely associated with the continental drift and plate tectonics. The unique features of the sub order Anthropoidea consists of huge brain with respect to their body size along, at the same time many other improvements such as time reduction in the apparatus for the sense of smell.

The shift of eyes in the forward area has been already done and the orbits are invariably detached from the temporal fossa, primarily because of extensive bony partition. The foramen magnum as well as the joint attached by the skull with the vertebral column lie forward during the early stages of development, but subsequently the orbits are, as it is expected, detached from the temporal fossa by extensive bony partition.

The permanent dentition mainly comprises of C=1 and I=2, M=2 or 3, PM=2 or 3, on each side of both the jaws which tend to drift underneath the brain case. The duration of the pre- and post-natal growth tends to extend to larger duration. The age of sexual maturity is already reached rather than being late in mostly all members, primarily in the higher simian primates.

The sub order Anthropoidea is generally categorized into three super family, on the basis of their distinct ecological niches. The three super families are the Platyrrhines/ Ceboidea all monkeys from America or New World Moneys, Catarrhines/Cercopithecoidea monkeys or Cercopithecoids from Africa and Asia also called the Hominidea and Old-World Monkey, including humans and apes across the world).

2. Tarsier

Tarsier, (family Tarsiidae), any of six or more than six species of small leaping primates found only in some of the islands of Southeast Asia, including the Philippines. Tarsiers are intermediate in form between monkeys and lemurs, measuring just about 9–16 cm (3.5–6 inches) long approximately, excluding a tail that is almost twice that length.

Tarsiers are lemur like in being nocturnal and are well known for its sense of smell. However, like apes, monkeys, and humans, the nose is hair-covered and dry, not bald and moist as is that of lemurs. The placenta and eyes are also simiiforme in structure.

The small brain of the tarsier has an enormous visual cortex in order to process information from the large goggling eyes which is the animal's most striking feature. The size of the eyes as well as visual cortex is made necessary by the absence of a reflective layer that is tapetum which the eyes of the most other nocturnal mammals possess.

The tarsier is also considered as unusual since they have especially long ankle bones named as tarsels and therefore, they are named as tarsier. They have shorter body and a round head that can be rotated 180 degrees. The face is short having large, membranous ears which are almost constantly in motion. The fur is thick and silly, and the color is grey to dark brown. The tail is kind of scaly on the underside similar to rat's tail. In a number of species, the tail is edging or terminal brush of hair.

Tarsiers are the only primates which are entirely carnivorous. They prey on insects, lizards and snakes. They cling upright to the trees and they press the tail against the trunk for support. The tips of their digits help in their grip because they expand into disk – like adhesive pads. Tarsiers move through the forest by going from trunk to trunk with the help of their greatly elongated hind limbs.

Adults do live in monogamous pairs and further keep in contact vocally at the night time while defending territory against other pairs by extremely high – pitched calls. These calls on the island of Celebes (Sulawesi) are the duets but complementary calls made by the male and female. The single – ones are bon in a state which is fairly well – developed furred and with the eyes open after a gestation of around six months.

Tarsiers are the ones who live on the islands of the southern Philippines, Celebs (Sulawesi), Bangka, the Natuna Islands, Borneo, Sumatra and Belitung. The species at these places differ from each other in such a way that there are some authorities which are inclined in order to classify them in different genera.

In the region of Indonesia and Malaysia, the Western tarsier that is Tarsius bancanus contains large eyes and thus, making the head broader than it is long. They contain longest feet and its tail is feathered at the tip. It benefits in both the old – growth as well as secondary forests. Despite this, it can also be found in the low scrubby vegetation and that too around the village. The South Sulawesi or spectral tarsier (T. tarsier which is formerly known as T. spectrum) is primitive, having smaller eyes, shorter feet, and further hairier tail.

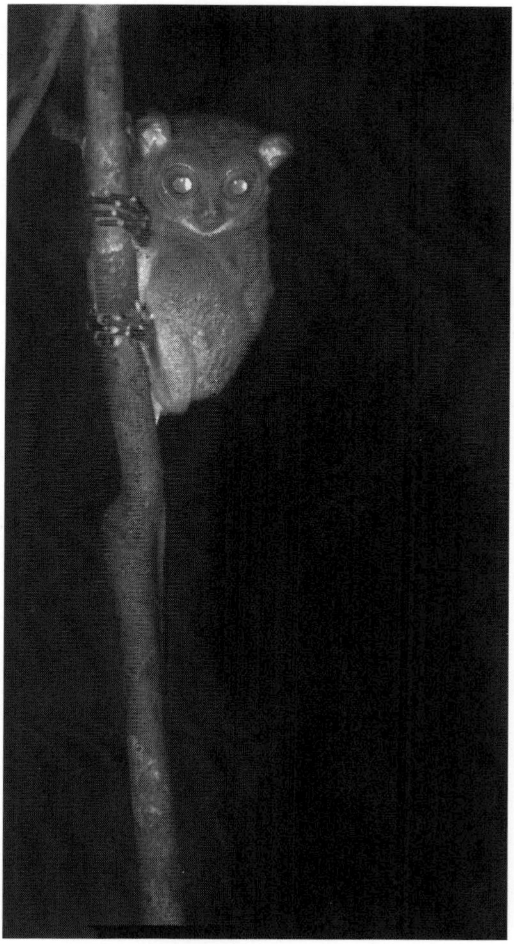

Figure 3.7. Western Tarsier.

Source: Image by Wikimedia

There are a number of various species on Celebes and its offshore islands but most of them are yet not been described in a scientific manner. The most distinctive is the high – mountain pygmy tarsier that is T. pumilus. The last living pygmy tarsier specimen was seen in 1921 until it was again seen in 2008.

The Philippine Tarsier that is T. syrichta contains a totally bald tail and their feet are also nearly hairless. Human settlement in its habitat makes them feel that their continued existence is in danger.in the suborder of

Haplorrhini, the family Tarsiidae is classified as monkeys, apes and humans but also it contains a separate infraorder that is Tarsiiformes.

New World Monkeys

There is a formal word which is used for the taxonomic group which contains all the New World Monkeys (NWMs) and that formal word is Platyrrhini. The New World Monkeys diverged from the Old-World Monkeys and Apes (Catarrhini) and this happed around 40 million years ago. It was thought that the ancestral species have migrated to the Americas on rafts of vegetation of by island hopping. The extant species of the New World Monkeys are classified into 5 families, 20 genera, 156 species and 204 taxa that is species and subspecies.

Table 3.1. A genus level classification of the extant New World monkeys

Family	Subfamily	Genus	Common Name/Synonyms	Spe-cies	Taxa
Atelidae	Alouattinae	Alouatta	Howler monkeys	12	19
	Atelinae	Ateles	Spider monkeys	7	15
		Brachyteles	Woolly spider monkeys/mu-riquis	2	2
		Lagothrix	Woolly monkeys	3	5
		Oreonax	Yellow-tailed woolly monkey	1	1
Aotidae		Aotus	Night monkeys/owl monkeys/ douroucoulis	11	13
Callitrichidae		Callibella	Roosmalens' dwarf marmoset	1	1
		Callimico	Goeldi's monkey/Goeldi's marmoset	1	1
		Callithrix	Marmosets	6	6
		Cebuella	Pygmy marmoset	1	2
		Leontopithecus	Lion tamarins	4	4
		Saguinus	Tamarins	20	34
		Mico	Marmosets	14	14
Cebidae	Cebinae	Cebus	Gracile capuchin monkeys	14	16
		Sapajus	Robust capuchin monkeys	8	9
	Saimiriinae	Saimiri	Squirrel monkeys	7	11
Pitheciidae	Callicebi-nae	Callicebus	Titi monkeys	31	31
	Pitheciinae	Cacajao	Uakaris	3	7
		Chiropotes	Bearded saki monkeys	5	5
		Pithecia	Saki monkeys	5	8

Source: https://www.nature.com/scitable/knowledge/library/new-world-monkeys-148121150/

The New World Monkeys exhibit striking diversity in their morphology, diet, behavior and distribution. Not so similar to the Old-World Monkeys, all the New World Monkeys are arboreal and with the exception of the owl monkey (Aoutus), diurnal. They are generally small to medium in size ranging from 0.1 kg to 15 kg. they are distinguished from the catarrhines by a number of various features.

Figure 3.8. Squirrel Monkey.

Source: Image by Pixabay

New World Monkeys have a broad and flat nostril whereas the catarrhines have narrow and pointing to downwards nostrils. New World Monkeys contains three premo9lar teeth whereas catarrhines have two. New World Monkeys have ear tube with bones (ectotympanic tube) which is also present in catarrhines.

Eventually, there is a difference in the configuration of the bones of the skull in between New World Monkeys and catarrhines. The New World Monkeys have the zygomatic and parietal bones which are in contact and the catarrhines have the frontal and sphenoid bones which are in contact. All the New World Monkeys have a tail which is prehensile in some taxa.

Prehensile tails have a naked patch of the skin in the direction of the ends for the reception of tactile stimuli which is proper for grasping, suspension and arboreal locomotion.

The New World Monkeys are found in Mexico, South America and Central America. Some countries contain extraordinarily rich primate diversity like the Brazil have 139 taxa from which 60 % are endemic and also Peru which have 50 taxa from which 14 % are endemic. They fill a large range of dietary niches which includes insectivory, gumnivory, frugivory and folivory.

Although, there are a large number of species which tend to consume a mixture of foods from various dietary categories. An alarming number of New World Monkeys are threatened with extinction and this is because of the consequences of habitat destruction and hunting like activities.

- Aotidae

The owl monkeys are sometimes known as douroucoulis and they consist of a single genus namely Aotus that contains 13 taxa. They all are distributed from Panama to northern Argentina. The Owl monkeys are the only nocturnal anthropoids, despite the fact that at least one species is cathemeral. They are basically small bodied having a round head and large eyes and inconspicuous ears. This word is derived from the Latin word 'a' which means without and 'Otis' that means ear.

Figure 3.9. Owl Monkeys.

Source: Image by Flickr

Their forelimbs are shorter in comparison to their hindlimbs and also, they have long bushy tails which helps them in balancing during the quadrupedal locomotion. Owl monkeys are primarily frugivorous, even though they consume leaves, insects and flowers. This genus is very unusual in being socially monogamous. They prefer to live in small groups of 4 to 6 individuals and consist of an adult pair, an infant, a juvenile and one or two subadults.

- Atelidae

The atelids are considered as the largest New World Monkeys which are of the weight 5 to 15 kg and thus, consists of two subfamilies such as Alouattinae and Atelinae. All of the five genera have a long and prehensile tails. In the Alouattinae, in which there is only one genus that is the howler monkeys named as Alouatta. The howler monkeys are different in their locomotion and morphology from the other atelids.

Figure 3.10. Red Howler Monkey.

Source: Image by Flickr

The number of the folivorous is more than most of the New World Monkeys and further minimize the energy expenditure through slow and deliberate quadrupedal locomotion and also very low levels of activity. They prefer to live in small groups having different species that exhibit different mating systems. Howler monkeys are among those which are the loudest terrestrial mammals and also, they have larynx which is highly modified along with a greatly enlarged hyoid bone which is in the shape of the cup.

It is thought to function as a resonating chamber for their calls. The subfamily Atelinae is much more energetic in comparison to Alouattinae and also, the atelines are very much frugivorous. Atelines tends to use their long limbs in order to travel rapidly over long distances by brachiation. They also contain some bigger brains in comparison to Alouttinae, which is relatable to their patchy and ephemeral fruit diet. There are four genera in the Atelinae.

- Callitrichidae

In the New World Monkeys, the marmosets and the tamarins are the smallest. This family consists of 7 genera as well as 62 taxa ranging from Panama to Southern Brazil. They are different from the other New World Monkeys with respect to their small size and claw like nails. They use their nails for vertical clinging on tree trunks.

They have much longer trunks, limbs and also tails. These are used for quadrupedal running and jumping. All of the callitrichids are frugivores and small animal predators. They also tend to consume plant exudates. Marmosets are specialist feeders and also, exhibit morphological adaptations in order to obtain, ferment and digest gums.

Figure 3.11. Common Marmoset.

Source: Image by Flickr

Callitrichids prefer to live in social groups which contain around 20 individuals. Whereas, marmosets prefer to live in larger groups in comparison to tamarins. Callitrichids are unusual in various aspects with respect to their mating and reproduction. Considering the mating system, they vary widely exhibiting monogamy, polygyny, polygynandry and polyandry.

Their groups often consist of extended family having a single breeding female and various breeding adult males. There are a number of groups which contain other non – breeding adult males and females who tend to help in caring for the infants. Usually, all the callitrichids give birth to twins.

- Cebidae

The family Cebidae consists of two subfamilies that is Cebinae that is Capuchin Monkeys and Saimiriinae that is squirrel monkeys which ranges from Central America to Argentina. Overall, there are 3 genera and 36 taxa. Their size is from small to medium weighing 1.5 to 5.0 kg and the capuchin monkeys are much larger in comparison to the squirrel monkeys weighing around 0.5 to 1.2 kg.

Figure 3.12. White-Headed Capuchin Monkey.

Source: Image by Pixabay

Cebids are further described as the arboreal quadrupeds. The tail in the capuchin monkeys is semi – prehensile and not in squirrel monkeys. Cebids are insectivore – frugivores. Despite this fact they consume a wide range of animal and plant foods. They have large brains for their body size, and this has been related to their complicated diets and relatively large home ranges.

Among New World Monkeys robust capuchin monkeys (Sapaius) that have been reported to be used as the tools during the foraging like cracking open palm nuts. Considering the social organization and mating systems, squirrel monkeys vary widely. Despite this they generally live in large groups having 15 to 75 individuals. Capuchin monkeys prefer to live in smaller groups consisting of 10 to 27 individuals.

In the poly-specific groups, the squirrel monkeys and capuchin monkeys are commonly observed together, and they can also include one or more other New World Monkey taxa. Such kind of associations can last hours, days or that can last even in years and so thought to increase foraging efficiency and thus, give better protection against predators.

- Pitheciidae

The pithecids shows up a diverse group of some not so usual looking and morphologically diverse New World Monkeys having equal unusual behavior and further ecological adaptations. The family is formed with 2 subfamilies, 4 genera and 51 taxa. Pithecids are predominantly herbivorous and along with some taxa which are specialized in the consumption of seeds (granivory).

Figure 3.13. Monkey Sakis.

Source: Image by Pixabay

They use quadrupedal locomotion and also lack prehensile tails. The size of the sakis (Chiropotes) varies from 1.5 to 3 kgs. They are elusive and also are known primates which live in small groups and further specialize on fruit and seeds. The bearded sakis (Pithecia) have extraordinary bouffant hair and there is a pronounced beard which is especially pronounced in males.

They have bigger body, weighing around 2 to 4 kg and are also specialized seed predators having wedge shaped canines and also powerful jaws. They prefer to live in larger groups having around 20 to 30 individuals. The body size and group size of the uakaris (Cacajao) which is same as the saki monkeys and they also have specialized teeth to intake seeds and hard fruits. Despite this, they differ in some aspects of morphology like the uakaris are the only New World Monkeys which have short tail.

- Old World Monkeys

The scientific name for this Old-World Monkeys is Cercopithecidae. Old World Monkeys (Cercopithecidae) are a group of siminians native to the Old-World regions which includes Africa, India and Southeast Asia. There are total 133 species in the Old-World Monkeys. This group include some members like macaques, geunons, talapoins, lutungs, surilis, snub – nosed monkeys, doucs, langurs and proboscis monkey.

Figure 3.14. Golden Snub-nosed Monkey.

Source: Image by Flickr

The size of the Old-World Monkeys varies from medium to large. There are some species which are arboreal while there are other which are terrestrial. The largest of all Old-World Monkeys is the mandrill and they weigh around 110 pounds. The smallest of the Old-World Monkey is the talapoin and they weigh around 3 pounds.

Generally, the Old-World Monkeys are stocky in build and also, have fore limbs which are in a number of species which are shorter in comparison to hind limbs. Their skull is heavily ridged and also, they have a long rostrum. Almost all the species are active in the day time and they also vary in their social behaviors.

There are a number of Old-World Monkey species that form small to medium sized groups with the complex social structure. The color of the fur of the Old-World Monkeys is grey or brown even though a few species have bright markings or even can have more colorful fur. The texture of the fur is neither silky nor wooly. In the Old-World Monkeys, the palms of their hands and soles of the feet are naked.

Diet:

There are a number of species of Old-World Monkeys which are omnivores and despite this, plants form the larger portion of their diet. Some groups are almost vegetarian, they eat leaves, fruit and flowers. Old World Monkeys also tend to eat insects, terrestrial snails and small vertebrates.

Classification:

Old World Monkeys are a group of primates. There are two subgroups of Old-World Monkeys named as Cercopithecinae and the Colobinae. The Cercopithecinae include primarily African species, for example mandrills, baboons, white-eyelid manga beys, crested mangabeys, macaques, guenons, and talapoins.

The Colobinae include mostly Asian species. Despite this that this group do include a few African species as well for example, black and white colobuses, red colobuses, langurs, lutungs, surilis doucs, and snub-nosed monkeys.

The species that come under the Cercopithecinae have cheek puches which is also known as buccal sacs which are used to store food. Since they have a varied diet, Cercopithecinae have non – specialized molars as well as large incisors. They contain simple stomachs. There are a number of species of Cercopithecinae which are terrestrial even though a few are arboreal.

In Cercopithecinae, the facial muscles are very well developed, and their facial expressions are used in order to communicate social behavior. The members of the Colobinae are folivorous and also, lack cheek pouches. They have complex stomachs.

Colobinae

Recently, the classification is done in order to propose three Africa and seven Asian genera. Throughout the African forests, arboreal colobus is found. In the west, the smallest as well as the least – studied genus that is Procolobus (olive colobus) which has the most restricted range and furthermore, it is represented by a single species.

Piliocolobus known as red colobus includes 16 allopatric species and is widely distributed from Senegal to Zanzibar. Many of them are endangered and one possibly extinct. The five most allopatric block and white Colobus species have the broadest distribution and also, occupy the widest variety of forest habitats.

Figure 3.15. Western Red Colobus Monkey.

Source: Image by Flickr

Similar to their African cousins, a number of Asian colobines are primarily arboreal, although the forests in which they live, cover a wider range of latitudes and altitudes ranging from lowland tropical to montane. The grey langurs (Semnopithecus) are the most terrestrial colobines which occur throughout the western Asian ranging from Sri Lanka to Nepal in habitat that varies from moist, scrub and conifer forest to the gardens as well as the temples.

Surilis (*Presbytis*) as well as lutongs (*Trachypithecus*) occupy dipterocarp forest and rubber plantations in Indonesia and Malaysia, with *Trachypithecus* species extending from north to China and west to India. Six allopatric taxa in the *Trachypithecus* Francoisi group, from which many are endangered, which live exclusively in limestone karst habitat, sleeping in caves and clambering on rocky cliffs.

Ranging and activity

Home ranges of most colobines are small that is less than 1 km^2, which reflects the relative abundance of the leaves that they eat. Even those with larger ranges typically travel less than 1 km per day. The enormous home ranges of Nepal gray langurs (*Semnopithecus schistaceus*) and snub-nosed monkeys (>10 km^2) reflect their relatively unproductive montane habitat. There are range sizes of many colobines that vary on the basis of season while tracking the proportion of different dietary components.

Social life

Colobine groups are generally moderate in size (10–20), which comprises of large number of adult females and very few adult males, which is1–2. Multi-male groups of some Asian colobines are usually recognized and known as traveling assemblages of one-male units.

Snub-nosed monkeys are astoundingly extreme — as 20 one-male units are capable enough to form bands of over 300 individuals (Kirkpatrick 2010). While on the other hand, other multi-male groups may only be temporary in nature, occurring at the time of male take-overs or when a son has reached the stage of maturity. Only the African red colobus is an exception to this pattern, as they are permanently living in larger multi-male groups of 25–50 (Fashing 2010).

Colobines mainly consists of various patterns of sex-biased dispersal (Fashing 2010, Kirkpatrick 2010). Male-biased dispersal characterizes African black-and-white colobus and Asian colobines, although it might be the case that females transfer groups; in some species, dispersing males, males that spread over wide area, join all-male groups.

In contrast, females are the ones that usually spread in African olive and red colobus. One of the noticeable exceptions in this is the endangered Zanzibar red colobus (*Piliocolobus kirkii*) that are living in human gardens at high densities, where philopatric females are more successful in accumulating food when supported by female kin (Siex 2003).

Figure 3.16. Zanzibar Red Colobus.

Source: Image by Flickr

The existence of one-male units in Asian colobines explains the stiff male-male competition for resident status. Indeed, several interactions between units might be aggressive in nature, and the species that are more consistent in associating units are the most sexually dimorphic (Grueter and van Schaik 2009).

The folivorous diet of colobines not often result in aggressive food competition between females, and between- encountered group primarily involve males, who seems to be protective mates. Unlike Asian colobine males, however, male ursine colobus (*C. vellerosus*) and guerezas (*Colobus guereza*;) seems to protect the food resources, perhaps as indirect mate defense. Indeed, female ursine colobus usually involved in evaluating males based on how well they fight (Teichroeb *et al.* 2009).

Infanticide, a dramatic expression of male-male competition, has been witnessed in many different colobines but is apparently absent in some species and population. Killing another male's offspring may be considered as an adaptive reproductive strategy in case infanticidal male is not the father and if it is likely that females will soon become sexually receptive than they would have otherwise and then mate with the infanticidal male.

It is also observed that strong support for this sexual-selection explanation comes from ursine colobus and gray langurs (*C. vellerosus;* Fashing 2010). In Thomas's langur (*Presbytis thomasi*), infanticide by extra-group males may result in inducing females that are without infants in a way to transfer with the infanticidal males to the smaller groups (Wich and Sterck 2010).

- Cercopithecinae

Though the available numbers of genera are easily comparable, cercopithecines are usually more widespread, ecologically flexible, and well-studied than colobines. They include both semi-terrestrial and arboreal forms and comprises of the largest variety of habitats of any major primate group. Out of the two cercopithecine tribes, *Papionini*, which includes baboons (*Papio*), macaques (*Macaca*), geladas (*Theropithecus gelada*), drills and mandrills (*Mandrillus*), and mangabeys (Lophocebus, Cercocebus, Rungwecebus), is generally studied in a more better way, especially those taxa that are common and more easy to notice, such as savannah baboons living in an open habitat semi-terrestrially. In contrast, it is sometime a challenge to follow mangabeys and the many guenons (*Allenopithecus*, *Cercopithecus*, *Miopithecus*) of the *Cercopithecini* tribe through their dense forest habitats (Cords in press).

Figure 3.17. Savanna baboon (Papio cynocephalus).

Source: Image by Flickr

It is usually seen that the species in the Forest dwelling area are more restricted with respect to geographic dispersion than those which are living in an open area or environments, although there are always exceptions, such as blue and DeBrazza's monkeys (C. neglectus, Cercopithecus mitis), which are distributed widely in sub-Saharan and geladas forests, that are limited to grassy, high plains in Ethiopia. It is often seen that whenever there are limited distributions of many forest-dwellers, resulted in making them vulnerable to hunting and habitat destruction; two such species —the one which is recently discovered Tanzanian kipunji (*Rungwecebus kipunji*) and the dryas monkey (*C. dryas)* of the central Congo basin — are critically endangered.

Macaques, which is the only Asian cercopithecines, are second only to *Cercopithecus* in terms of number of species and to *Homo* in geographic range. Barbary macaques (*Macaca sylvanus*), their African representatives, are the only non-human primates, that are in existence, north of the Sahara region, and a recently introduced European population lives in Gibraltar. Asian macaques are spread over a large area from Japan to India, occupying habitats varies in nature with respect to mountains, lowland rainforests, cities, and temples. Just like the colobines, most of the cercopithecines are tropical, but several macaques occur at higher, sometimes snowy elevations, including rhesus (*M. mulatta*), Barbary, Taiwanese macaques (*M. cyclopsis*), and Japanese (*M. fuscata*).

Ranging and activity

It is important to note that the Cercopithecines have the ability to travel farther than colobines, as they need to travel long way in a way to search for food. They also get very less time to rest and more time in socializing, as well as grooming as a major form of social exchange. It is also seen that most of the forest-dwelling species form mixed groups of species, which may function in predator defense. These associations are more common among mangabeys and guenons, less so among sympatric colobines and macaques in the same forest.

Social life

It was observed that the group size and composition are highly variable across and within cercopithecine species (Cords in press). Usually the group consists of 10–100 individuals, but in certain instances, it may be beyond 1200 individuals, and mandrill 'hordes, which are persistent social units, approximate 620 members (Swedell 2010).

The groups of most of the papionins comprises of large number of males, but some, such as geladas, hamadryas baboons (*Papio hamadryas*), and, possibly, drills (*Mandrillus leucophaeus*) as well as guinea baboons (*P. papio*), have consist of smaller one-male groups as basal social units. These units are categorized into a hierarchical system of troops and bands. Single, multi-male and less frequent mangabey groups merge into 'supergroups.'

Such sub-grouping is considered as one of the best ways to ensure or maintain flexibility in a way to effectively address the rapidly changing ecological conditions and to negotiate shared resources such as sleeping cliffs and watering holes. Mandrills possess of unique type of social flexibility — that are not common among primates, their massive hordes do not carry adult males outside of their breeding season (Abernethy *et al.* 2002).

In contrast to most of the papionins, majority of the guenon groups include only one adult male for most of the year. During their seasons of breeding, however, resident males may be overwhelmed by the large number of mate-seeking bachelors. It is usually seen that the non-resident cercopithecine males are solitary in nature; only in Japanese macaques, geladas, and some guenons do there exist all-male groups that are comparable to those of Asian colobines (Cords in press).

Mostly, all cercopithecine females exists in their natal groups for their whole life; only one exception in this case is hamadryas baboons. Females form the kin-directed spatial proximity, stable social core, coalitionary and grooming support are common. It often results in development of stable matriline-based hierarchies, but dominance style varies from species to species.

'Despotic' species, such as Japanese macaques, rhesus, vervets, and baboons, exhibit steep and strict female dominance relations, with little reconciliation and strong kin biases. In contrast, three Sulawesi macaques (*M. tonkeana*, *M. nigra*, and *M. maurus*) demonstrate a more symmetrical 'egalitarian' dominance style (Thierry 2010).

The members of the *Cercopithecini* tribe, however, not always in accordance with the traditional cercopithecine pattern: female patas monkeys, though they are philopatric, very less frequently form stable, matrilineally based hierarchies, and there is only limited evidence available of blue monkeys that female rank is in correlation with the proxy of biological fitness (Foerster *et al.* 2011). It is always observed that the female blue monkeys nether form frequent coalitions; instead they may simply reduce contests over food by ignoring each other while feeding.

Figure 3.18. Blue Monkey (Cercopithecus mitis stuhlmanni).

Source: Image by Wikimedia

Relations that are found in between male cercopithecines are often antagonistic. While on the other hand, grooming is rare, though some species of the male form coalitions in order to improve their chances of success.

Irrespective of the fact that they are being kin-biased — males, being the dispersing sex, are not closely associated with each other — their preference to participate in coalitions seems to depend, instead on mutual advantage or reciprocity, in conjunction with the differential bargaining positions.

It is usually seen that females typically mate with more than one male, during the time of seeking extra-group partners, and may involve or indulge in sexual activity even when they cannot conceive. These types of behavior of females often confuse paternity, thus averting infanticide, which always happens in many cercopithecines.

Sexual swellings, copulation calls, and color changes signaling ovulation may incite male-male fighting, mate guarding and/or sperm competition. Females of most of the species have preferences for individual males, and not likely the highest-ranking ones.

On the other hand, in case of multi-male groups, the extent to which females can select mate choice depends on different number of factors such as degree of sexual dimorphism, breeding synchrony, and habitat (i.e., the feasibility of hidden rendezvous). In case of groups of one-male guenon, influxes of extra-group males may cause more difficult for resident males to monopolize breeding.

Table 3.2. Primate Order Taxonomy (only major species groups are included in this table)

Suborder	Infraorder	Parvorder	Superfamily	Family	Subfamily	Common Names	Distribution
Strepsirrhini (strepsirrhines: primates with moist noses.) (also known as prosimians)	**Lemuriformes** (lemurs)			Lemuridae (true lemurs)		ring-tailed, and ruffed lemurs	Madagascar and Comoro Islands (for some species)
				Lepilemuridae		sportive lemurs	
				Cheirogaleidae (dwarf lemurs)		dwarf and mouse lemurs	
				Indriidae		indris, avahis, and sifakas	Madagascar
				Daubentoniidae		aye-ayes	
	Lorisiformes (lorises)			Loridae		lorises, pottos, and angwantibos	lorises--India and Southeast Asia; others--Africa
				Galagonidae		galagos (or bush babies)	sub-Saharan Africa and Zanzibar
Haplorrhini (haplorrhines: primates with dry noses)	**Tarsiiformes** (tarsiers)			Tarsiidae		tarsiers	Philippines, Borneo, Celebes Islands, and Sumatra
	Simiformes (anthropoids)	**Platyrrhini** (New World monkeys)		**Callitricidae**	Calitricinae	marmosets and tamarins	Panama and north and eastern South America

				Cebidae	Cebinae	squirrel and capuchin monkeys	forested areas of Central America and/or north and eastern South America
					Aotinae	night and titi monkeys	
					Atelinae	howler and spider monkeys	
					Pithecinae	uakaris and sakis	
		Catarrhini (Old World monkeys, apes and humans)	**Cercopithe-coidea** (Old World monkeys)		**Cercopith-ecinae**	guenons, vervets, baboons, macaques, etc.	guenons and baboons--Africa; macaques--northwest Africa, Gibraltar, South and East Asia
					Colobinae	colobuses, langurs, and proboscis monkeys	colobuses--Central Africa; langurs--India and South-east Asia; proboscis monkeys--Borneo
			Hominoidea (apes and humans)	**Hylobatidae** (lesser apes)		gibbons and siamangs	Southeast Asia
				Hominidae (hominids--great apes and humans)	**Ponginae**	orangutans	Sumatra and Borneo
					Gorillinae	gorillas	Central and West Africa
					Homininae	chimpan-zees and bonobos	Central and West Africa

						humans	originally only tropical tropical and subtropical regions of Africa

Source: https://www2.palomar.edu/anthro/primate/table_primates.htm

REFERENCES

1. https://www.biology-online.org/dictionary/Haplorrhine [Accessed 27 Aug. 2019].
2. https://www.livescience.com/55276-lemurs.html [Accessed 27 Aug. 2019].
3. https://www.nature.com/scitable/knowledge/library/new-world-monkeys-148121150/ [Accessed 27 Aug. 2019].
4. http://cec.nic.in/wpresources/module/Anthropology/PaperII/4/content/downloads/file1.pdf [Accessed 27 Aug. 2019].
5. https://onlinelibrary.wiley.com/doi/pdf/10.1002/9781119179313.wbprim0045 [Accessed 27 Aug. 2019].
6. https://www.thoughtco.com/old-world-monkeys-130648 [Accessed 27 Aug. 2019].
7. https://www.nature.com/scitable/knowledge/library/old-world-monkeys-83033815/ [Accessed 27 Aug. 2019].
8. https://science.jrank.org/pages/5517/Prosimians.html [Accessed 27 Aug. 2019].

Chapter 4

Distribution and Diversity of Primates

"Everything is improved by the judicious application of primates."

— Chris Roberson

CONTENTS

This chapter emphasizes on the distribution and the diversity of the primates all around the globe. Primates shows a remarkable range of size and adaptive diversity all over the world. There exist few species that have extended their living range outside the tropics. The living primate species consists of an astonishing variety of appearances, sizes and specializations so that one considers it unbelievable at times that we all have a common ancestor. Also, there exist natural groups of extant primates. There exist five major natural groups of primates on Planet Earth. These groups are further classified on the basis of their habitat and living habits.

4.1. INTRODUCTION

The placental mammals belong to the order primate. This is considered as one of the most diverse groups of mammals. This group of mammals is having three hundred or more species and six hundred or more subspecies.

The species such as Apes and monkeys are also known as non-human primates. These species are distributed in the Paleotropical (old world) and neotropical (new world) biogeographical zone. This group of mammal depicts a broader adaptive diversity with their remarkable range of body size.

They are distributed all over the tropical latitude at elevation up to 5000m of msl. In addition to it, few species all over the world live at moderate and cold mountain zone (for instance, Rhinopithecus species) where temperature goes below zero degree.

The Asian primate includes thirteen genera and one hundred and seventy-six species & subspecies. India has seventeen species of primates and out of these species ten species are harbored at the region of north east, the Indo-Myanmar hotspot region.

All these species are critically endangered. These species are endangered and vulnerable because of the severe ecological as well as anthropogenic threats.

This work presents data on the relative diversity, abundance, and distribution patterns of primates in a 20 km2 region of the Tapiche River in the Peruvian Amazon. The data of the population was collected while the study area was both inundated as well as dry (March to September 1997). The population data was collected by the use of conventional line-transect census techniques.

Figure 4.1. There is a wide distribution of the primates all over the world.

Source: Image by Pxhere

The outcomes of the Survey reflected the existence of eleven primate species, but population parameters on only eight of the species will be presented, comprising saddleback tamarins (Saguinus fuscicollis), Bolivian squirrel monkeys (Saimiri boliviensis), brown capuchins (Cebus apella), white-fronted capuchins (Cebus albifrons), monk sakis (Pithecia monachus), red titi monkeys (Callicebus cupreus), red uakaris (Cacajao calvus), and red howler monkeys (Alouatta seniculus).

In addition to it, woolly monkeys (Lagothrix lagotricha), night monkeys (Aotus nancymaae) as well as pygmy marmosets (Callithrix pygmaea) were also seen in the region. The data for the smaller-bodied primates is similar to that which was reported nearly eighteen years earlier. On the other hand, the data for the larger-bodied primates reflect a loss in the number of animals that are existing in the area.

Pressure from hunters as well as the timber industry may account for decreasing the numbers of large-bodied primates, while it looks as if the natural features peculiar to the conservation area make contribution to the patchy pattern of distribution.

If one takes a look at the living primate species, one come across an astonishing variety of appearances, sized and specializations so that one at times considers it unbelievable that we all have a common ancestor. This common ancestor originated near about eighty million years ago when the so-called Primatomorpha fragmented in the ancestors of today's colugos (Dermoptera) and the first primates.

Therefore, the closest relatives of the primates are not, as long believed the Treeshrews (Scandentia), but the characteristic of cat-sized mammals is a membranous structure that covers all its extremities and the tail that enables the animals to slide from tree to tree (Mason et al 2016).

While only two colugos species are recognized in the Southeast region of Asia, over five hundred primate species from sixteen diverse families are known. The number of species has increased quickly in the past twenty years. This is because of the new possibilities of species differentiation using genetic methods.

Traditionally, the geographic gradients in the species richness of non-human primates have been attributed to the variation in forest productivity (associated to precipitation levels), though an all-inclusive, global-scale analysis has never been done.

There were many scientists or experts who performed a more comprehensive test on the role of precipitation as well as biomass production and suggest an alternative hypothesis, the variation in vertical structure of forest habitats as calculated by forest canopy height - in determining the primate species richness on a worldwide scale.

Considering the potential causal relationships among precipitation, forest structure and productivity, the experts arranged these variables inside a path framework to measure their direct and indirect connections with the pattern of primate species richness using structural equation modelling.

In addition to it, the analysis accounted for the impact of spatial autocorrelation in the relationships and measured possible historical differences between the biogeographical areas. The path coefficients show that the forest canopy height (which is used as a proxy for vertical forest structure) is a better predictor of primate species richness than either precipitation or productivity on both worldwide as well as continental scales.

The only exception was the region of Asia, where precipitation prevailed, albeit independently create productivity or forest structure. The impact of spatially structured processes differed markedly amongst biogeographical regions.

The results challenge the traditional rainfall-based viewpoint in favor of forest distribution and structure as primary drivers of primate species richness, which aggregate potential impacts from both climatic factors as well as habitat complexity. These discoveries may support predictions of the influence of removal of forest on primate species richness.

4.2. SIZE RANGE AND DIVERSITY IN PRIMATES

Primates represent a remarkable range of size and adaptive diversity. Madame Berthe's mouse lemur (Microcebus berthae) of Madagascar is the smallest primate. The weight of the primate is somewhat 35 grams. Primates are widely distributed all over the tropical latitudes. These primates are living at elevation up to 5000 m. It has been observed that few species have extended their range outside the tropics, the moderate region. Barbary ape (Macaca sylvanus) living in temperate of the Atlas mountain.

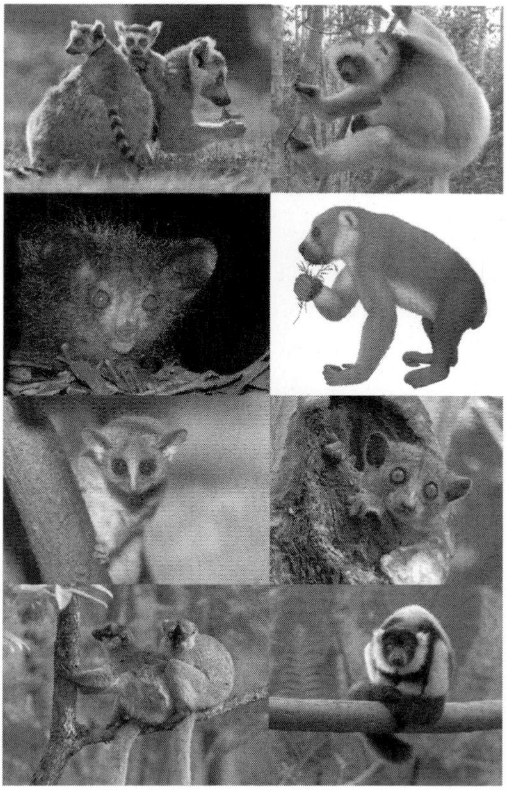

Figure 4.2. Primates all over the world depicts a notable range of size and adaptive diversity.

Source: Image by Wikimedia

Tibetan macaque and Japanese macaque live in Cold Mountain. Snub nose monkey (Rhinopithecus species) lives in the region of China where temperature drops below zero degree. Biogeographically, Primate species which exists in Paleotropical region are called as old-world monkey. On the other hand, the primate species of Neotropical region are known as new world monkeys.

There are some continents like United States of America, Australia, European countries and Middle east countries and Russian where non-human primates are completely absent. The existence of primates is evident in many of the countries such as Neotropical countries, African countries, South Asian countries comprising China, Japan and Madagascar only.

The regions like Neotropical region and African region are considered as the most primate predominant regions. It is described based on species area relationship as these two regions (Neotropical region and African region) have greater land mass amongst the primate habitat countries. Asian primates occupy third place with thirteen genera and one hundred and seventy-six species amongst primate habitat countries. Asian Primates richness is because of the two confounding factors.

Members of the order Primates show a remarkable range of size and adaptive diversity. The most massive primate is certainly the gorilla (Gorilla gorilla). The weight may be over 4,000 times as great, changing from 140 to 180 kg (near about 300 to 400 pounds).

Primates occupy two main vegetational zones. These are tropical forest and woodland–grassland vegetational complexes. Each of these zones has produced in its resident primates the suitable adaptations, but there is possibly more diversity of bodily form amongst forest-living species than among savanna inhabitants.

One of the descriptions of this alteration is that it is the precise pattern of locomotion rather than just the simple matter of habitat that rules overt bodily adaptations. Within the region of forest there are several numbers of ways of moving about. For example, an animal can exist or live on the forest floor or in the canopy. Also, within the canopy the animal can move in three particular ways.

These ways involve:

- by leaping—a function principally dictated by the hind limbs;
- by arm swinging (brachiation)—a function mainly of the forelimbs;

- by quadrupedalism—a function which is equally divided between the forelimbs and the hind limbs. On the savanna, or in the woodland-savanna biome, which considerably demands for adaptations for ground-living locomotion rather than those for tree-living, the likelihoods are limited. If bipedal humans are discounted, there exist a single pattern of ground-living locomotion, which is known as quadrupedalism.

There are at least two variations on the theme within this category:

(a) knuckle-walking quadrupedalism, and

(b) digitigrade quadrupedalism.

The former gait is considered as a characteristic of the African apes (chimpanzee as well as gorilla), and the latter of baboons and macaques, which walk on the flats of their fingers. After human beings, Old World monkeys of the subfamily Cercopithecinae are considered as the most successful or effective colonizers of nonarboreal habitats.

Primates are considered as a homogeneous group morphologically. It is only in the realm of behavior that the variations or changes among primate taxa are clearly discriminant.

It can be stated that the most successful primates (judged in the context of the usual criteria of population numbers and territorial spread) are those primates that have deviated least from the ancestral pattern of structure but farthest from the ancestral pattern of behaviour.

4.3. NATURAL GROUPS OF EXTANT PRIMATES

From the viewpoint of geographical distribution, five 'natural groups' of extant primates are apparent, which offers a platform for all groupings and evolutionary trees. The initial group covers the lemurs (Lemuriformes). Lemurs are the only primates that are present on the island of Madagascar, where they underwent a main adaptive radiation.

In addition to the almost eighty species of lemur that is alive today, that array encompassed near about twenty mostly larger-bodied sub-fossil species — some of the species are almost as big as a gorilla — that died out during the last two thousand years or so, after the first humans colonized Madagascar.

The inclusive ranges of the five natural groups: lemurs, lorisiforms, tarsiers, New World monkeys, Old World monkeys and apes. Lorisiforms occupy most of the regions but not all of the region inhabited by Old World monkeys as well as apes.

The second natural group comprise lorises, pottos and bushbabies (Lorisiformes). They are far more extensively spread, occurring in Africa and region of South and Southeast Asia. But their adaptive radiation was more limited, including only some thirty species.

Figure 4.3. Lemurs is a single indigenous species that is found on the island of Madagascar.

Source: Image by Wikimedia

The third group, comprising a dozen or so generally alike tarsier species (Tarsiiformes). This represents an even smaller array in the Southeast Asian archipelago.

The fourth natural group, comprising the monkeys of South and Central America (Platyrrhini). They are considered as the only extant primates that is found in the New World, where they diversified widely, yielding some one-hundred-and-thirty known species. On the other hand, Platyrrhine monkeys fall into two fairly distinct groups, the relatively small, tamarins, claw-bearing marmosets and the larger-bodied true monkeys.

The fifth group, comprising Old World monkeys and apes (infra-order Catarrhini). This group is the largest as compared to all the other groups. This group with over one hundred-and-fifty species in South Asia, Africa and Southeast Asia.

In spite of extensive overlap in geographical distribution, monkeys as well as apes of the Old World falls in to two different sister groups. Humans also count among the ape subgroup of catarrhine primates but extended far beyond the original geographical range during the later phases of their evolution.

The first three natural groups of primates (lemurs, loris group, tarsiers) are often collectively called prosimians (literally 'pre-monkeys') or 'lower primates' because they have generally retained a greater proportion of primitive characters.

Prosimians are all comparatively small, with a modal body weight near around 500 g, and most of them are active at night. In fact, several lines of evidence show that ancestral primates, such as the ancestral placental mammals, were nocturnal.

As a rule, primates are clearly either nocturnal or the primates are diurnal. But certain lemurs, notably bamboo and brown lemurs, signifies a highly unusual blend of nocturnal and diurnal activity, for which Ian Tattersall coined the term 'cathemerality'.

Contrasting with prosimians, the 'higher primates' or simians — apes, monkeys and humans — share a distinguishing suite of advanced anatomical features in their jaws, teeth, brain and reproductive system. All simians are diurnal except for the owl monkeys of the New World.

They are typically considered as large-bodied, with a modal weight of near about 5 kg. Across primates in general, body sizes represent a three-thousand-fold range from 40 g in the case of pygmy mouse lemur to 120 kg in case of an adult male gorilla.

Even though they have retained many primitive characteristics, tarsiers represent various specific resemblances to simians in olfaction, vision, brain morphology and reproduction. This has long recommended an ancestral link between tarsiers and simians that has been confirmed by multiple lines of molecular evidence.

Because of this reason, another basic difference is drawn between lemurs and lorisiforms on one hand and tarsiers and simians on the other hand. Lemurs and lorisiforms are called as 'strepsirrhines'. This is because they have retained a naked, moist area of skin (rhinarium) around the nostrils. Tarsiers and simians are known as 'haplorhines' because, uniquely amongst mammals, they have secondarily lost the rhinarium.

4.4. DISTRIBUTION AND ABUNDANCE

The nonhuman primates have an extensive distribution in the entire tropical latitudes of India, Africa, Southeast Asia, and South America. Within this tropical belt, which lies between latitudes 25° N and 30° S. The nonhuman primates have a considerable vertical range within the tropical belt which lies between latitudes 25° N and 30° S.

In Ethiopia the gelada (genus Theropithecus) is found existing at the elevation point of up to 5,000 metres (16,000 feet). Gorillas of the Virunga Mountains are known to travel across mountain passes at altitudes of over 4,200 metres at the time of traveling from one high valley to the other.

Figure 4.4. It has been observed that few species of nonhuman primates are found in the exterior of the tropic.

Source: Image by Wikimedia

The howler monkeys of Venezuela (Alouatta seniculus) is found living at the elevation point of upto 2,500 metres in the Cordillera de Merida. On the other hand, in northern Colombia the durukuli (genus Aotus) is found in the tropical montane forests of the Cordillera Central.

In habitat, primates are predominantly tropical, but there are few species of nonhuman primates that extend their ranges well outside the tropics. The Barbary "ape" (Macaca sylvanus) are found in the temperate forests of the Atlas and other mountain ranges of Morocco and Algeria.

There are some populations of rhesus monkey (M. mulatta) extended until the middle of the 20th century to the latitude of Beijing in the northern region of China. In addition to it, Tibetan macaque (M. thibetana) is found from the warm coastal ranges of Fujian (Fukien) province to the cold mountains of Sichuan (Szechwan).

One of the most remarkable, though, is the Japanese macaque (M. fuscata), which in the north of Honshu is found in mountains that are covered with snow for eight months of the year. Some of the populations of rhesus monkey have learned to make life more bearable or acceptable for themselves by spending most of the day in the hot springs that bubble out and form pools in volcanic areas.

Finally, two western Chinese species of snub-nosed monkey, the golden (Rhinopithecus roxellana) and black (R. bieti), are confined to high altitudes (that is up to 3,000 metres in the case of the former and to 4,500 metres in the latter). Here the temperature falls below 0 °C (32 °F) every single night and often barely increases above it by the daytime.

Although there exist many primates that are still plentiful in the wild, the numbers of many species are decreasing suddenly. According to the International Union for Conservation of Nature (IUCN), over seventy percent of primates in the region of Asia and roughly forty percent of primates in the region of South America, on the island of Madagascar, in mainland Africa are listed as endangered.

It has been observed that a number of species, mainly the orangutan, some of the Madagascan lemurs, the gorilla and some South American species, are in serious danger of extinction unless and until their habitats can be conserved in perpetuity and human predation is kept under control.

The populations of various species number only in the hundreds. In the year 2000 a subspecies of African red colobus monkey (Procolobus badius) became the first primate since the year 1800 to be stated extinct.

The populations of some critically endangered primate species have increased in the middle of these declines. Concerted attempts to breed a kind of marmoset, the golden lion marmoset (or golden lion tamarin; Leontopithecus rosalia), in captivity have been effective; reintroduction of that species into the wild remains in the region of Brazil.

The projected number of western lowland gorillas (G. gorilla gorilla), a species that is thought to be critically endangered, increased when a population of over 100,000 was explored in the year 2008 in the swamps of the Lac Télé Community Reserve in the Republic of the Congo.

4.4.1. Social Distribution of Primates

One of the most important theoretical contributions made by studying primates is an understanding of what impacts the nature of complex variation in social structure as well as organizations. Highlighting this is a set of theoretical developments investigating why animals live or survive in groups in the first place and why the groups are of a particular size. In this scenario, the main emphasis is on the question of "what impacts the size of groups that the primates choose to live in?"

Particularly, primates are useful taxa in order to address this question because their group sizes are highly variable, not just within, but also among the species. Intraspecific group size for red colobus (Procolobus rufomitratus), for instance, differs between twelve and one hundred and fifty members (Chapman & Chapman 2000a).

As seen in orangutans, interspecific variation ranges from a single individual, to more than eight hundred members in mandrills (Mandrillus sphinx). In addition to it, within some species, social groups split and re-unite repeatedly into subgroups of diverse sizes and combinations with time (for instance, chimpanzees (Pan troglodytes), and fission-fusion social organization of spider monkeys (Ateles sp.), or the multi-level organization of gelada (Theropithecus gelada) as well as Hamadryas baboons (Papio hamadryas, Aureli et al. 2008).

This interspecific variation provides the foundation for investigators or scientist to develop models in order to research both the ecological as well as the social drivers of group size.

Grouping is helpful in various number of ways. Individuals in bigger groups are thought to have a reduced risk of predation, may be better able to find and protect food resources, and may be protected against conspecific threat, such as infanticide by extra-group males.

There were many investigators who have recommended that grouping confers such predictable advantages that differences which is present in the group size can be explained by the disadvantages or shortcomings. The most extensively accepted potential cost of grouping is thought to be a decrease in foraging effectiveness.

Being with other individuals which have the similar dietary needs means that animals either fight over food (that is contest competition), or one animal present in a group beats another animal to the food. Therefore, when the second animal comes to an area there is simply no food left for the animal to eat.

In both of these scenarios, it is thought that competition over food which forces the animals to travel farther. The logic behind this statement is comparatively simple. Animals must forage over an area that can fulfil their energetic as well as nutritional needs.

It follows that a rise in the size of the group will increase the area that must be covered in order to find adequate food supplies. Therefore, people must travel further and spend more amount of energy if they are in a bigger group, as compared to those in a smaller group.

With the rise in the time that is spent in traveling, a point is approached where the energy that is spent in travel is too expensive and smaller groups become beneficial. In this way ecological factors can have impact on the movement patterns as well as foraging efficiency, thereby constraining the size of groups that can effectively exploit the food resources that re available.

The important component of the ecological constraints model is that the increase in the size of the group must give rise to an increase in within-group feeding competition. It is believable that this functions in a slightly different fashion whi relies on the nature of the resources that are used by particular species.

With frugivorous, and possibly many folivorous primates, that feed in discrete patches, typically trees bearing food items or clumps of trees — additional members of the group may reduce the patches quickly. And, this give rise to increased day ranges.

On the other hand, for more insectivorous species, whose resources may not occur in as discrete patches, continual travel all over the canopy in search of insects with other group members may give rise to an increase in the intersection of individual search fields, decreasing per capita encounter rates with food and therefore, increasing the area that must be examined.

Therefore, the ecological constraints model recommends that those factors or aspects that have impact on the distance that animals travel should also have the impact on the size of the group. For those animals that typically feed on fruit or leaves that can be exhausted, the size of the patch would predict for what duration a group of a given size could stay and feed.

A large group would expend less time in a patch of a given size as compared to that of a smaller group. This reason behind this is that it depletes the patch faster. A large group simply has a greater number of mouths to feed. If animals travel between the patches once they have depleted them, then the density as well as distribution of patches will predict the travel costs that is inflicted.

The distance that is needed to travel to the next patch is small, travel costs that are inflicted are low, and animals can therefore form large groups in case when the resource patches are at a high density or in a clumped distribution.

At such times, any additional cost that is related with being a member of a large group, such as the requirement to visit many patches, can be recovered in an easy manner. In contrary to it, when resource patches occur at low densities, the distance that is needed to travel to the next patch is typically large, travel costs that is inflicted are high, and animals cannot afford to quickly deplete patches, and thus animals form small groups.

It is important to note that the fission-fusion social organization of chimpanzees, spider monkeys, and some other primates provides valuable tests of the ecological constraints model because with this, it is possible to assess the temporal changes in subgroup size from direct assessment of the density, size, and distribution of food resources (Chapman 1990, Chapman et al. 1995). For chimpanzees and spider monkeys, a unique and helpful regression model was developed to explain that the density, size, and distribution of food patches assessed and calculated every month were a good predictor of subgroup size.

It is often complicated to visually characterize a multiple regression, but the selected methods that were used for spider monkeys are explained and described in the next few sessions. It is usually observed that the spider monkeys have a highly flexible fission-fusion type of social organization. in addition, in the Santa Rosa National Park, Costa Rica, the range of the subgroup size varies from 1 to 35 individuals, but on an average basis, almost 5 individuals are often found traveling together. Out of this population, it is easy to predict mean variance of almost 50% of the monthly subgroup size from relatively crude measures of the density, size, and distribution of food patches (Chapman 1990).

In addition to the species that are discussed above, the application of this ecological constraints model has been supported by the research and number of studies coming from a variety of species (Snaith & Chapman 2005) and situations (e.g., general models, Wrangham et al. 1993, and mixed species associations, Chapman & Chapman 2000c). Given the increase in the importance that this model has been receiving, the time has come when there is a need to suggest that species should rely or stand upon the expectations of the ecological constraints model.

However, it does not mean that all species will conform, and thus it becomes even more important to engage in the research activity in this field because investigators possess the ability to search for exceptions to the model. It is assumed that those species that do not conform will do so because they got familiar with or adopted social strategies that run counter to the ecological expectations. This gives a valid reason to evaluate or assess the potential importance of different and multiple social strategies (i.e., the further one deviates from the expectations of the ecological model, the more important the social selective pressures).

4.5. SURVIVING CONDITIONS FOR PRIMATES

Primates are a taxonomic group for which distribution patterns and diversity are relatively more understood. Primate diversity is primarily dependent on different number of environmental drivers in distinct regions and locations, but a detailed global analysis of these drivers is not yet available or done. Regional patterns of the richness of primate species have been attributed, for instance, to the disparity in precipitation rates, either directly or as a substitution for productivity.

While the level of precipitation levels has been directly associated to the richness of species through tolerance optima, the productivity hypothesis proposes that areas with high level of productivity that helps in supplying resources for more individuals and thus, to even more species (Wright 1983). In fact, climate should also limit the distribution activity of non-human primates at high latitudes, explaining the beginning of clade under conditions which are comparable to those found in present-day tropical regions. Within these tropical (as well as adjacent subtropical) boundaries, however, richness of species is dependent on number of related factors other than productivity that also covary with precipitation.

The diversity in the vertical forest structure, for instance, has been directly associated to both plant productivity (as primarily dependent on the amount of rainfall; Woodward 1987) and the incidence angle of solar radiation, thus explaining and showing a clear latitudinal gradient. Owing to the close association of primates with tropical forests, the distribution and structure of these forests is in fact a good indicator of the diversity of the microhabitats or habitats available for primates, with different number of species being found in structurally more complex forests.

By contrast, the simple structure of forest – through habitat fragmentation, for instance – cause negative impact on the overall population or community.

In addition to giving more importance to the precipitation-based hypotheses, earlier models of richness patterns of the primate species have mainly focused on the records of local assemblages, which are comparatively more vulnerable to the problems of area effect, sampling parameterization, and missing species. Many of these studies have often ignored the effects of spatial autocorrelation and related inference problems.

Given these existing limitations, the primary focus here is on a comprehensive, global data set of environmental variables and primate distributions in order to reassess the past productivity-based hypotheses and test an alternative one – vertical forest structure – as rival predictors of primate species richness on an international scale, while measuring for only spatial autocorrelation in the data.

Specifically, the best possible way is to employ a path-analysis framework to contrast both the dependent and independent associations of environmental productivity, precipitation, and forest vertical structure with primate species richness. These studies are of high importance for the researchers as it helped then in gaining more insight into this. The researchers also assessed and evaluated these relationships within the principal biogeographical regions that are valuable for primates, in a way to examine or identify the possible historical differences in these relationships.

4.6. ARE HUMANS DESCENDANTS OF APES?

In the present times, for scientifically educated humans, it goes without question that all living beings are a product of evolution and that a common ancestor for all living primates comprising humans existed. Therefore, the issue, which is to find the common ancestors, seems to be of semantic nature.

The scientifically suitable term would be "primate". In German-speaking science, the Anthropoidea, the group that has divided from the Tarsier after dividing of the "dry-nosed" primates, is referred to as the "real monkey" and in recent time frame, present literature, only as "monkeys". It comprises of all families of Old-World monkeys, and the New World, including apes - and thus humans as well. In case if one denoted to the first anthropoids that lived almost 64 million years ago as a monkey, one must wisely presume that our ancestors of that period, around 64 million years ago, were monkeys. However, this question has only one reasonable answer: not only humans are descendants of primates but is one too.

It is not at all controversial to picture humans within the context of the placental mammal group that humans belong to, the primates. Neither it is infrequent for the anthropologists, primatologists, or biologists of other sorts to compare the anatomy, sexual or social or behavior, cognitive abilities or lifestyles of humans with those of other primates.

Irrespective of this, Homo sapiens is mostly ignored or missed out when authors assess or examine the living primates of the world, a perhaps common mistake as works refers to or about 'other' animals, but a misleading one given that it give more weight to view lineage as part of the group; as a conspicuous and primary part of primate diversity.

- 'Human' as explained in this entire section denotes Homo sapiens alone. When H. sapiens and its extinct relatives are being referred to, the term 'hominin' is used.
- Almost all books that covers the topic of primate diversity do include these species within the roster: Walker's Mammals of the World does, so does Redmond's The Primate Family Tree and Dunbar and Barrett's Cousins: Our Primate Relatives, for example.

In many respects, Homo sapiens is an 'extreme' primate. The cultural diversity, vocal abilities and application of technology (which includes the use of fire) encompasses that of other primates, and it can be assumed that it is the most advanced and unparalleled technological innovation which directed us to adopt to particular lifestyles – those involving livestock management, agriculture, or regular travel at sea, for instance, that are so different with respect to those of other primate species.

In most of the cases, it is not unusual at all and fit within the wider scope of primate diversity: our substantial cultural and anatomical diversity is what that can be predicted for a widespread primate that exists in large number of the habitats, many of the behavioral and anatomical traits that once appeared to be unique to humans are not unique at all, and the present existed social systems are applicable within the spectrum present in related primate species.

It is interesting to note and observe new findings at a time when we compared oneself with our close relatives. The more knowledge and understanding about primate biology, evolution, and history, the lesser the gaps become between Homo sapiens and other related primate species, and the less unusual in appearance. Indeed, most of basic knowledge and traditional understanding about human 'uniqueness' certify the fact that the species that result in bridging the gap between human beings and their closest living relatives are extinct.

As it is currently not widely discussed topic that that humans are unusual apes in that human beings appear more likely how the monkeys look. In fact, people appearance just like the monkeys helps in explaining the idea of a 'pre-pongid' divergence of the human lineage that was widely recognized among anatomists prior to the molecular age.

The basic understanding or belief that living (non-human) apes are all somewhat specialized and thus completely different from species that were their ancestral to the human lineage has often been mentioned. With the discovery of more fossil apes, it become clearer that hominids and hominid-like hominoids carrying a diversity of body plans, some of which appear like monkeys than 'ape-like'.

However, by evaluating the fossils, it can be assumed that the hominins evolved from among a radiation of species with 'ape-like' proportions; the belief that the common ancestor of the whole African ape clade was Bonobo-like overall has been proposed, and the present available data from australopithecines justifies that some, at least, had 'ape-like' long forelimbs and especially the short hindlimbs ones, though they didn't possess the same skeletal specializations for knuckle-walking or for climbing as extant African apes.

The evolution of especially forelimbs and of long hindlimbs are more in proportionate of monkeys as compared to apes. Therefore, the innovations of hominin evolved somewhere on the branch between the earliest members of Homo and certain australopithecine species.

4.6.1. Let's get it out of the way: apes are monkeys

Having mentioned monkeys, there's something we have to get out of the way. The widely common belief that humans are apes is accepted by many, especially by those people who follow tree-based thinking. Furthermore, despite the traditional belief that humans are 'different enough' from other primates to warrant their own group, currently it is believed that Australopithecus, Homo, as well as their close relatives should be united with gorillas, chimps, and other great apes in a group that come under the category of single common ancestor.

This group is Hominidae, within which the African ape branch is termed Homininae (the hominines), within which the human branch is termed Hominini (the hominins). Along with their 'lesser ape' relatives the siamangs and gibbons (the hylobatids), hominids belong to the more include clade called Hominoidea.

So, humans are considered apes from the perspective of phylogenetic point. That is, in scientific parlance, 'ape' no longer known or recognized as 'non-human hominoid' (critics of this might make a point that, in common parlance, 'ape' does indeed mean 'non-human hominoid', but common parlance does not dictate best practice). But, for a minute if one believes that humans are apes, then it's time to bite the bullet and acknowledge that both humans as well as apes were also once monkeys. Again, there are some obscurity that monkeys are small, possess tail, and do not include apes but, from a tree-based, phylogenetic perspective, apes belong to the group of large, tailless monkeys.

Most of the people are still opposed to this idea, proving their point that the term 'monkey' should be limited to the non-ape anthropoids (Anthropoidea is the clade that comprises of both New World monkeys and both apes and Old-World monkeys). But if whales are artiodactyls, birds are dinosaurs, and humans are apes then, dammit, apes are monkeys.

4.7. PRIMATES IN COMMUNITIES

Community ecologists have been long investigating the interaction and community structure since the early 20th century (e.g., Clements 1916). However, in recent times, these topics have been investigated in primates more specifically as the traditional primatologists have tended to give more importance to the population ecology. Nonetheless, there has been increase in the knowledge and research of competition, parasite-host, predator-prey, and plant-primate interactions (e.g., seed dispersal and herbivory), as well as biogeographic differences in community structure. In the next few paragraphs, readers gain an insight into competitive, predatory, parasitic, and mutualistic interactions that primates have with other species — both plant and animal.

It has been observed that most of the non-human primate communities are more or less closed to contact with the members which belongs to the other communities. Most frequently, they are tied to a specific locale and they rarely travel to the exterior of their home range. This aloofness from other troops or the communities prevents high concentrations of individuals. This could result in quick depletion of local resources.

Chimpanzees are considered as a prominent exception. When chimpanzees from the diverse troops come together, there is often an exciting, friendly encounter which lasts for several hours, following which, some of the adult females switch groups. Apparently, they are searching for new mates.

Occasionally, however, contact or interactions between communities of the comparatively unpredictable chimpanzees will grow into genocidal violence.

Contact or interactions inside the communities of non-human primate are usually unlimited. Subgroups are hardly closed from group interaction. All members of a community have face to face, casual communication on the daily basis. The most common kind of subgroup comprises a mother and her young offspring.

In some forest living primates, contact or interactions among the groups of the similar species is in the form of a specialized territorial defense behavior. Instead of avoiding each other, the members of the groups actively meet near their common territorial border and make hostile displays.

For instance, Howler monkeys, siamangs, indris, and gibbons all produce exceptionally loud vocalizations for this purpose. All four of these species live in home ranges that are generally so small that the food resources of bordering territories can be seen and become attractive.

4.7.1. Competitive Interactions

It is usually seen that within a community, all species including primates have their own or possess unique niche. However, it is very complicated to understand and define, a niche can be defined as particular place or position of an organism in a community and the ecological roles that it performs in the same community. Hutchinson (1957) explain a more detailed definition that identifies a niche in n-dimensional hypervolume in which these dimensions comprises of the environmental conditions that define a degree up to which these species can survive.

The fundamental niche refers to the complete range of environmental conditions in which a species can survive; a realized niche refers to the smaller set of environmental conditions that a species exploits in the presence of competition. It is generally observed that the chances of the competition (contest or scramble) is highest among closely related, sympatric species that mainly possess similar niche space.

Interspecific competition is even higher or stiff when the availability of the resources is limited or either temporally or spatially scarce and/or highly favored as a function of their quality. When it is expected that extant species will compete vigorously for the critical resources, then there are higher chances that one or more of the interacting species may witnessed population decline or character displacement in a process known as competitive exclusion.

As the primate species are competing with each other for the food resources, it results in demanding attraction among the researchers to gain an insight into this, especially with respect to female-female competition, male-male competition, and inter-group competition. However, these are population level interactions.

It is very difficult to understand and explain the interspecific competition, especially with species that are long-lived such as primates. In short, in order to demonstrate that one species is out competing another species, it should be explained that the presence of the bigger competitor is negatively impacting the microevolution or abundance of the other species.

Suggesting that local abundance and species' adaptations are a function of past interspecific competition makes several assumptions. The first assumption is that the environment is constant over time, and that the variables shaping interactions in the current time frame are the same variables that were important in the evolutionary and ecological past ("Ghost of Competition Past", see Connell 1980).

The second assumption is that the population growth in species a is determined by species b, when in fact there are many other non-competition related factors that directly affect the population abundance (e.g., resource availability, predation, disease). There is no confusion that interspecific competition over the food resources play very important role in shaping community interactions and assemblages.

However, the complication in explaining this empirically has cause many researchers to assess or evaluate species coexistence, which instead of theorizing past competition, focuses on the mechanisms and processes of allowing extant species to live sympatrically in the ecological present.

REFERENCES

1. https://www.ncbi.nlm.nih.gov/pubmed/11376449 [Accessed 27 Aug. 2019].
2. https://journals.sagepub.com/doi/full/10.1177/1940082916680104 [Accessed 27 Aug. 2019].
3. https://www.nature.com/scitable/knowledge/library/what-influences-the-size-of-groups-in-58068275/ [Accessed 27 Aug. 2019].
4. https://www.researchgate.net/profile/Anwaruddin_Choudhury2/publication/265405344_Primates_in_Northeast_India_An_Overview_of_their_Distribution_and_Conservation_Status/links/5715da1108ae1a840265078f.pdf [Accessed 27 Aug. 2019].
5. https://www.britannica.com/animal/primate-mammal/General-considerations [Accessed 27 Aug. 2019].
6. https://www.degruyter.com/view/j/mamm.2018.82.issue-6/mammalia-2016-0136/mammalia-2016-0136.xml [Accessed 27 Aug. 2019].
7. Wolfheim, J. (1983). *Primates of the world.* Seattle: University of Washington Press.
8. https://wwnorton.com/college/anthro/evolve6/ch/05/review.aspx [Accessed 27 Aug. 2019].
9. http://anthropology.msu.edu/iss220-fs12/files/2012/08/understanding_humans_ch06.pdf [Accessed 27 Aug. 2019].
10. https://onlinelibrary.wiley.com/doi/abs/10.1002/ajp.1017 [Accessed 27 Aug. 2019].
11. https://besjournals.onlinelibrary.wiley.com/doi/full/10.1111/1365-2656.12241 [Accessed 27 Aug. 2019].
12. https://www.ncbi.nlm.nih.gov/pubmed/19430965 [Accessed 27 Aug. 2019].
13. https://www.researchgate.net/publication/282629896_Primates_Distribution_Status_And_Threats [Accessed 27 Aug. 2019].
14. https://www.jstor.org/stable/10.1086/286078?seq=1#page_scan_tab_contents [Accessed 27 Aug. 2019].
15. https://www.dpz.eu/en/info-center/knowledge/primates/evolution-and-diversity-of-primates.html [Accessed 27 Aug. 2019].

16. https://blogs.scientificamerican.com/tetrapod-zoology/why-humans-are-important-to-studies-of-primate-diversity/ [Accessed 27 Aug. 2019].
17. https://bioone.org/journals/primate-conservation/volume-2014/issue-28/052.028.0103/Diversity-Abundance-and-Habitats-of-the-Primates-in-the-R%C3%ADo/10.1896/052.028.0103.full [Accessed 27 Aug. 2019].
18. https://www.ncbi.nlm.nih.gov/pubmed/24773500 [Accessed 27 Aug. 2019].
19. https://www.researchgate.net/publication/228782720_Distribution_and_Diversity_of_Primates_in_Guyana_Species-Area_Relationships_and_Riverine_Barriers [Accessed 27 Aug. 2019].
20. https://www.sciencedirect.com/science/article/pii/S0960982212008068 [Accessed 27 Aug. 2019].
21. https://www2.palomar.edu/anthro/behavior/behave_2.htm[Accessed 27 Aug. 2019].
22. https://www.nature.com/scitable/knowledge/library/primates-in-communities-the-ecology-of-competitive-59119961/ [Accessed 27 Aug. 2019].

Chapter 5 Interaction between Primates and Other Species

"Interestingly, bonobo percussionists prefer a tempo of 280 beats per minute, the syllabic rate at which most humans speak"

— Dr Susan Block

CONTENTS

The primates are animals that prefer to live in groups and interact with each other. This chapter provides an overview of the subsets of the species in the community and the food webs in the community in the first half. There is an interaction in between the predator and the prey which has been described in this chapter. It mentions the primates as predators and primates as prey and so, gives explanation regarding the same. This chapter provides the insights of the interaction between chimpanzee and non – human species and the social structure of the primates. Primates as a natural heritage and primates' social cognition has been explained in the second half of the chapter.

5.1. INTRODUCTION

Defining and Measuring Communities

It is a well-established fact that no species can thrive in isolation. Every single living life form live and connect with different species and are vital individuals of an ecological community. The configuration of species inside a community and the connections among them in both is affected by long - term (evolutionary time scale) and short - term (ecological time scale) processes.

Longer-term transformative and biogeographic procedures shape the traits pertinent to species association with each other, pattern of distribution, the variety of the species, while momentary natural interactions can impact how intense the interaction could be. It also influences the amount of species present through the span of a season, a year, or within the lifetime of a living being (Morin 1999, Verhoef and Morin 2010).

5.1.1. Subsets of species in a community: Taxocenes & Guilds

There is abundant diversity of life forms on the planet (a significant number of which have not been found at this point). It is not possible to investigate all the species in network. Constantly, society group environmentalists study subsets of species inside a group. A Taxocenes, for instance, is a subset of species sharing a common taxonomic classification.

The monkeys and gorillas of Kibale National Park is a case of a Taxocenes: they are classified inside the primate order (Infraorder Catarrhini) and live in a similar forest in Uganda. The guild is described as a subset of species inside a community that utilizes a lot of resources in the same manner. The prosimian primates, bats, and insects are an example of a guild that is a nocturnal pollinating guild; these animal species are not closely associated taxonomically, however are sympatric and suck the nectar of night-blossoming blooms in similar manner.

5.1.2. Food Webs in a Community

The way species will interact with other species is determined by the position it occupies in a food web. The position is a main factor. Fundamentally, food webs are the sum of all predator-prey contact at work and are a depiction of how energy moves among species. The fundamental trophic level contains plants (autotrophs) that utilizes solar energy by converting carbon dioxide into organic compounds, especially sugars, in a methodology known as photosynthesis. Thus, the plants are defined as producers since they are the reason whereupon heterotrophic living things exist, the organisms recognized as consumers. The second trophic level is occupied by Primary consumers and are plant consuming animals that is herbivores. The third trophic level is occupied by secondary consumers,

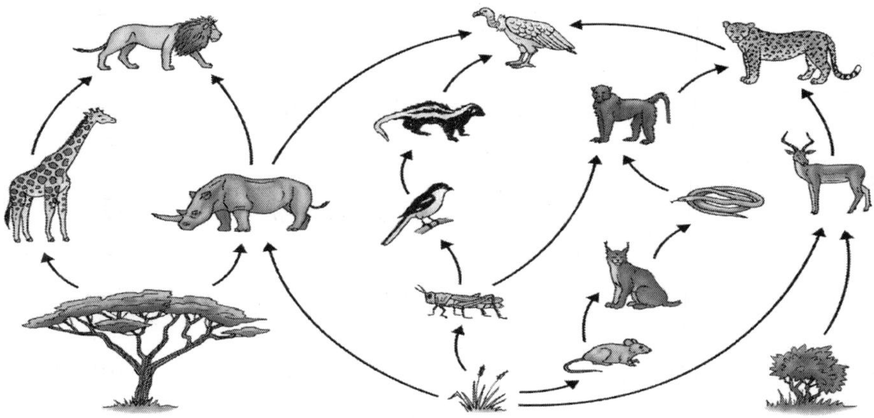

Figure 5.1. The Food Web.

Sources: Image by Flickr

They eat herbivores and are called carnivores. The carnivores can similarly eat carnivores and occupy the fourth trophic level, and so forth. Animals that consume food from more than one trophic level are called omnivores. The majority of primates are omnivores, regardless of the way that there are several groups of primates that have variations for pure herbivory (e.g., Colobinae, Alouatta spp.) or carnivory (e.g., Tarsius spp.).

5.1.3 Species Interactions in a Community

The communities are best studied as a group of species because they are linked to each other intricately in the food webs. These affiliations will have either a negative or positive repercussion for all of the two or more species, and subsequently have basic implications for the ecology, adaptations, and evolutionary history of all creatures. Since at any rate two species are incorporated, a contact can be described in context of the plusses and minuses in the participation of those two species. For example, predation is an association wherein one species takes advantage in light of the fact that it has acquired its dinner, while various species is clearly oppositely influenced because it is the dinner.

5.1.4. Primates in Systems

Figure 5.2. The primates in a Group.

Sources: Image by Pixabay

The various scientists have inspected species contacts and system structure since the starting of twentieth century (e.g., Clements 1916). Though, these subjects have quite recently been looked into in primates even more expressly as historically primatologists have would when all is said in done highlight population ecology. Regardless, we are continuously getting information into competition, predator-prey, parasite-host, and plant-primate interactions (e.g., herbivory, seed dispersal), similarly as biogeographic complexities in community structure (Fleagle et al. 1999).

5.1.5. Competitive Interactions

All the species together with primates have special place inside the system. Though it is not easy to describe, a niche can be normally investigated as the positioning of a living being in a system and the ecological roles it plays in that system. The Hutchinson in the year of 1957 gave an undeniably clear definition that recognizes a niche in n-dimensional hypervolume in which the dimensions are the environmental conditions that describe the range in which species kind can survive.

The various environmental conditions in which a species can exist is defined as fundamental niche; a narrower set of environmental situations that a species uses in the competition exists is defined as realized niche. The possibility for competition (scramble or contest) is very much at peak among closely associated, sympatric species that reside on same niche space. At the point of time when existing species are observed or expected to compete directly for essential resources, one or more of the interacting species are expected to experience population decline or character displacement in a process known as competitive exclusion according to Gause in 1934.

That primates compete with each other is well known and has attracted great research, particularly concerning male-male competition, female-female competition, and between intergroup contention. These are people level coordinated efforts, regardless. The competition that is interspecific test is significantly much difficult to represent, especially with long living species, for instance, primates. Essentially, to show that one species is winning over another species, it must be exhibited that the closeness of the superior competitor is oppositely influencing the abundance or microevolution of various species. To begin with, it is expected that the environment is constant over time, and that the variables shaping relationship in the present are same variables that were significant in the ecological and evolutionary past (Ghost of Competition Past)

5.2. PREDATOR-PREY INTERACTIONS

5.2.1. Primates as predators

The primates can be both predator and prey species. As predators, primates go after both plants (herbivory) and animals (carnivory). The angiosperms diversified throughout the Mesozoic Era; thus, mammals and different living beings emanated into a large group of plant-devouring specialties.

According to Potts in 1996, about 255 million years back herbivory advanced and there was a drastic change from past periods that did not have primary consumers.

In the modern mammal orders the primates are ones that have developed to sustain basically on plants. Even omnivorous primates eat plants to meet a large share of their nutritional and energetic necessities. There are some primates whose diet are entirely made out of leaves or seeds and other plant parts, for example, howler monkeys (Alouatta spp.), mountain gorillas (Gorilla berengei), leaf-eating monkeys (colobinae), and saki monkeys (pitheciinae).

According to Lambert in 1998 these species have developed characteristics of the dentition and gut that support their obligate Folivory (leaf eating) or granivory (seed eating). As per Norconk in 1998, Pitheciine monkeys are specific seed predators and have dental anatomy and expertise to open tough hard, unripe fruit to utilize under-developed seeds; in this job, pitheciine are associate of a seed predator organization that contains parrots. According to Lambert in 1998 Colobine monkeys have developed stomachs that can ferment plant fiber and can thrive on a plant diet, for example the ruminant ungulates. The Folivorous primates can possibly strip trees of their leaves and in turn cause their death. For instance, Glander in 1975 reported due to the regular consumption of an Alouatta palliata group in Costa Rica the mortality of trees was high.

Figure 5.3. Monkey Consuming Leaf.

Sources: Image by Pixabay

Folivorous primates can exert considerable leaf predation pressure, especially in societies in which leaf-eating monkeys are found in large numbers irrespective of the fact that they don't slaughter the tree totally. Struhsaker in 1997 reported that generally a normal of 356 individual colobines living in a square kilometer, and each one of them eat plants and leaves, almost throughout the day in Kibale National Park.

The primates generally consume creatures, majority of that are invertebrates especially insects and little vertebrates that contains reptiles, amphibians, birds, and eggs of birds. There are some bigger bodied primates for example, Papio anubis, Pan troglodytes that can eat bigger vertebrates, for example, forest duiker, impala, bushbuck, bush pigs, and even different primates, for example, galagos (Galago spp.). In any case, it is the chimpanzee (Pan troglodytes) that has gotten most focus as a predator animal variety. The other primates will in general be divided as opportunistic predators, chimpanzees give off an impression of being exceptionally sorted out and hunt by cooperating with each other.

5.2.2. Primates as Prey

According to Isbell in 1994 the large majority which is 73 percent of primate species live in social groups. As group living expands the potential for intragroup feeding competition, if there was no benefit to being social, it would not have been so common. One theory is that the primates live in society because of the danger of predation. Generally, all the primates are prey species, that includes the biggest bodied primates — gorillas and chimpanzees. According to Jaffe and Isbell in 2010 both the terrestrial and aerial predators consume primates as they are prey species. That includes cats for example, leopards, Panthera pardus, domesticated dogs (Canis lupus familiaris), fossa (Cryptoprocta ferox), snakes (e.g., pythons, Python sebae), monitor lizards (Varanus niloticus), and the variety of raptors. The African crowned hawk eagle in Africa (Stephanoaetus coronatus) is a predator and has been known as a primate specialist predator.

According to Mitani in 2001in Kibale National Park, Uganda, 83 to 88 percent of this eagle species diet consist of monkeys.

According to Isbell and Young in 1993 the leopards that is Panthera pardus can genuinely affect primate population; they have been accounted for to consume 14 individuals from a single vervet monkey (Cercopithecus aethiops) social group in just 30 days in Amboseli National Park, Kenya. The primates are also hunted by humans; in reality, commercial hunting (bushmeat business) is an extraordinarily difficult issue putting in danger primates in South America, Asia, and especially Africa.

Figure 5.4. Turkey Vulture Eating Snake.

Sources: Image by Wikimedia

5.2.3. Parasite-Host Interactions

The parasite host contact plainly has negative consequences for one variety, and a positive result for the other in the same way as predator-prey communication. There are many parasites that can feed on primates easily. The health of host does not have much effect due to some parasite host contact, while others can affect fitness and life span of the host. Ectoparasites are other parasites that can be found in the fur, skin, and superficial flesh of their hosts.

The neotropical botfly is the general primate ectoparasite, that can put eggs in meat to consume it by growing larvae. They form into huge flies if it is not treated; scratching can prompt skin diseases that might be responsible for high death among howler monkeys as per Milton in 1996. In the body of the host the parasite like Endoparasites are found, most regularly in the gastrointestinal tract, and also in blood, lungs, and other body tissues and organs.

The host's parasite burden can differ by individual, season, and area. For instance, in both the neo-and paleo-tropics, there is a higher occurrence of parasitic contaminations and furthermore more chances for re-disease during rainy seasons. According to Stuart and Strier in 1995 when parasites can harm host, they are a possibly powerful selective force on primates and can support behaviours that decrease danger of becoming infected. For instance, it is known in certain species (e.g., Alouatta palliata) that animals will even more regularly change places of sleeping during periods of high parasite occurrence according to Milton in 1996.

5.2.4. Polyspecific Associations

The primates take part in contact that are of shared advantage to every single participating species even though the contact can have some bad effects on species like competition, predation, and parasitism. Some primate species, for instance, form polyspecific relationship in which social group of at least two primate species aggregate. By coincidence the formation of some relationships take place, when two social groups of various species happen to meet up in a similar place during their day by day tasks. These relationships are generally found in regions with high population densities as well as between monkey species utilizing common food place. Others though appear to be nonrandom.

5.3. CHIMPANZEE INTERACTIONS WITH NONHUMAN SPECIES IN AN ANTHROPOGENIC HABITAT

There are large number of complicated paths in which the contact happens between the different natural life species. The cross-species connections can be persistent or speedy and rare, extending from commensalism to predation. There are some cross-species communications in mammals, like the, predation or anti predatory relationship between species for defensive reasons well researched and known. According to Endler in 1991, most different types of connection go unreported or keep on being anecdotal, for example, circumstances like cross-species play or alloparenting in nature.

Instances of such examples include a female Japanese macaque (Macaca fuscata) driving on the lower back of sympatric deer (Cervus nippon yakushimae) in Yakushima Island in Japan (Kiyono non-public communication), a wild capuchin monkey (Cebus libidinosus) nursing a marmoset according to Izar in 2006, or a lone lioness in the Samburu Game Reserve in Kenya adopting and protecting oryx (Oryx beisa) calves according to Douglas-Hamilton in 2002. This latter example is fabulous as it concerns a huge bodied flesh eater and its customary prey species.

The research on cross-species contacts can permit us in a best way learn about animal behavioral and cognitive flexibility and how male and female species can in a best way or adversely affect different species around them. Attributes, for example, body size, behavioral adaptations, intellectual capacity and level of interest likely influence the species inclination to start interactions, notwithstanding site-specific points, for example, flora and flora in large numbers or territory type.

Figure 5.5. Primate interacting with others.

Sources: Image by Flickr

In particular the large bodied animals can possibly be less willing to escape from different animals according to Cooper and Stankowich in 2010 and might therefore, be more possible to initiate interactions with smaller species. Omnivorous species will in general show extra varied interplay types than carnivores, as they are less plausible to consider different species prey and to be seen as a danger. In conclusion, cognitively advanced species may be anticipated to classify distinctive species according to Matsuzawa in 2006 and display more prominent interest and adaptability as to sympatric wildlife according to Mitani in 2010.

Chimpanzees (Pan troglodytes) provide a best model to find out patterns of cross-species associations. While they are large-bodied, they are in most circumstances ripe fruit specialists. However, seize and eat a variety of mammals, birds and different vertebrates according to Wrangham in 1977; Nishida in 1979; Goodall in 1986.

The difference in behaviour that is population or focused on communities have been reported across a range of animal taxa, but the level of differences across social and tool-use domains singles out the chimpanzee according to Whiten in 1999; McGrew in 2004; Waal in 2005; Sanz in 2009.

Narrative perceptions prescribe the presence of several communication types among chimpanzees and different species, and adaptability in chimpanzee reactions to different species according to Teleki in 1973. In any case, most connections between wild chimpanzees and sympatric vertebrate species are agonistic, and most research till this date has given attention on chimpanzee predatory behaviour. The societies of Chimpanzee and individuals contrast in predation statements and hunting behaviour, routinely actively paying searching for the prey and luring in every cooperative and opportunistic hunt according to Newton-Fisher in 2007.

The accessibility of prey, visibility of prey, prey portability, and seasonality affects communities' searching patterns (Sugiyama, 1989; Uehara; Ihobe, 1998; Gilby., 2006; Gilby; Connor, 2010). Chimpanzee prey profiles are consistently broad, varying from birds to various primates (Wrangham & van Zinnicq Bergmann-Riss, 1990; Uehara et al., 1992). All things considered, however, chimpanzees all through Africa prey in the primary upon colobine monkeys, particularly purple colobus, Procolobus spp. (Boesch & Boesch, 1989; Stanford et al., 1994; Uehara, 1997; Mitani & Watts, 2001).

According to Hosaka in 2001 the diverse variables of prey behavioral ecology, for example, activity patterns and speed of movement might also influence communication charges (Boesch & Boesch, 1989). There are regions in which the required prey species are not found, in which case chimpanzees might also aim at other animals they do not normally eat, for example, galagos (Galago spp.), mongooses (Herpestidae family) and pangolins (Manis spp.) utilizing present day techniques to get right of entry to the prey, for example galago spearing in Fongoli, Senegal; Pruetz & Bertolani, 2007).

The Chimpanzees mainly living in territories of anthropogenic effect eventually at some point of tropical Africa, and previous research exhibits chimpanzees designate conduct methodologies to make the strategies to make the most forest-farm habitats (Hockings et al., 2012; Hockings & McLennan, 2012). The significantly affected anthropogenic habitats generally offer less wildlife species. In spite of the fact that data are rare, in such natural surroundings chimpanzees might also divide animals in another manner to their counterparts in significantly less disturbed environments with greater availability of wanted prey species.

In human-affected territories the Chimpanzee population residing may respond to pet species with indifference, aggression or with predatory expectations. For instance, chimpanzees at Gombe in Tanzania forcefully go after a domestic cat (Felis domesticus) according to Teleki in 1973, and chimpanzees in Uganda have been said to kill domesticated species, in particular chickens according to McLennan in 2010.

Chimpanzees frequently react aggressively to domestic dogs (McLennan & Hill, 2010). This particular species can be a hazard to chimpanzee protection in anthropogenic habitats according to McLennan, 2010. Chimpanzee contact with different species are no longer restricted to predatory context, and though less regular, they may also interact with flora and fauna in different manner.

The young chimpanzees have been found to interface in fun loving behaviour with various vertebrate species (e.g., duiker, Cephalopus sp., Boesch & Boesch, 1989; flying squirrel, Anomalurus derbianus, Boesch & Boesch, 2000), even despite the fact that these same species can generally become their prey (e.g., primates, Papio anubis, Goodall, 1986).

Besides, chimpanzees might also trap and kill animals. Beside eating the animal meat, they may choose to play with the carcass (e.g., hyrax (Dendrohyrax dorsalis), Hirata et al., 2001). The monitoring of wild and hostage chimpanzees suggest that non-eating communications are by and large done with the support of adolescents or teenagers (Teleki, 1973;

Goodall, 1986; Boesch & Boesch, 2000; Ross et al., 2009).

According to Sabater Pi, in 1993 he observed bonobos (Pan paniscus) from the Lilungu area of the Democratic Republic of Congo (DRC) taking part in intimate, non-predatory and long-lasting social endeavor with more younger colobus monkeys that they catch. According to Ihobe in 1990 in the same way at Wamba in DRC it is observed that adult male colobus monkeys prepared bonobos in an affiliative or fun loving style, while guenons (Cercopithecus ascanius; C. wolfi) regularly moved toward bonobos yet did now not start direct contact, whereas travelled, eat and rested in near areas.

Additionally, the environment has been affected by humans and its expansion. researches that assess how creatures alter their behavior ecology to all elements of anthropogenic regions are essential for identifying species' behavioral flexibility according to Hockings in 2012.

Such information will moreover highlight the extent to which particular species can adjust to ecological changes.

5.4. SOCIAL STRUCTURE

In large social groups or societies is where the majority of primates live and that includes humans. In the situations of semi-terrestrial primates' varieties such as baboons, living in a large group gives safety from predatory cats, dogs, and hyenas. The rare food resources are also secured in this manner. This is particularly correct for non-human primates when fruit is a food. Leaf-eaters, for example, colobus monkeys and langurs, will in general come together to form small social groups as there is small challenge for their food. The primates that are nocturnal are generally small and usually hunt on its own.

There are special cases to these general examples. For example, the veggie lover orangutans normally live alone aside from moms with their kids before they go off leaving their families.

The majority of non-human primate communities remain pretty much in isolation and do not form contact with primates from different societies. Regularly, they are attached to a specific region and seldom migrate outside of their home range. This lack of approachability from different troops restricts high number of individuals which could bring about quick depletion of natural resources. The external species are generally not accepted, and the communities are normally aggressive towards them.

Therefore, social communications between individuals from different groups are normally uncommon, particularly for females. Chimpanzees are a prominent special case. At the time when chimpanzees from various groups meet up, there is generally an exciting, friendly encounter which continues for few hours, after that some of the group females change the groups or go in another group as they are looking for new mates for the purpose of reproduction. There is a chance of large-scale violence during the time of contact between communities of chimpanzees with unpredictable behavior.

Connections inside non-human primate networks are normally unlimited. The Subgroups are rarely closed for group interaction. All individuals from a community have every day eye to eye, casual contact. The most widely recognized kind of subgroup comprises of a mother and her young child.

The contact between groups of similar species is in form of a specific territorial defense behaviour in some primates living in forest. The groups generally assemble close to their common territorial border and make unfriendly displays instead of neglecting each other.

5.4.1. Non-human Primate Social Group Composition

In the primate's social groups where there is large scale variation between them, there is almost no variation within each species. The majority of non-human primate species are restricted to just one of the following six basic patterns

- single female and her child
- monogamous family group
- polyandrous family group
- one-male-several-female group
- multimale-multifemale group
- fission-fusion society

1. Single Female and Her Offspring

This phenomenon is normal in most primates, but the single female and her child group pattern is uncommon for some. It happens in the orangutans and a few of the small nocturnal prosimians (example mouse lemurs and galagos). The grown-up monkeys spend their lives generally alone. In any case, they meet up with females at times for mating. Both male and female children generally leave their mother when they experience sexual development.

2. Monogamous Family Group

The adult male and female with their children consists of monogamous groups. When they are developed, the youngsters leave to make their very own family units. Though this group pattern is the most well-known one for humans, it is uncommon for non-human primates. It is found among the small Asian apes and in some of New World monkeys and prosimians. In particular, monogamous family groups are the common pattern for gibbons, siamangs, titi monkeys, indris, tarsiers, and obviously some pottos.

3. Polyandrous Family Group

The marmosets and tamarins which are the smallest New World monkeys, build both monogamous and polyandrous nuclear families. They by and large begin as a monogamous mating pair. Afterward, a second adult male monkey may join the family and support in rearing of child. Both the adult males will conceivably mate with the adult female as soon as it happens. This type of practice is common as these monkeys generally have twins and the dads bear the children on their back's majority of the time. This polyandrous mating example is incredibly uncommon among non-human primates yet occurs in some human societies in isolated village areas of India, Sri Lanka, and particularly Nepal, and Tibet.

4. One-Male-Several-Female Group

This type of group has polygynous mating patterns. In other words, one male normally mates and lives with more than one female at once. Polygyny is commonly not an immoral mating design. In place of this, the male and his female mates structure a separate mating and child rearing group. This example is found in hamadryas primates, geladas, langurs, howler monkeys, gorillas and numerous human societies. It has been a socially favored marriage pattern in various Native American, African, and South Asian societies. However, polygyny is not as normal among people as monogamy, even in regions that support it.

It would be an error to assume that in a non-human primate one-male-a few female gatherings are commanded by males. Between geladas, females to a great extent control the social group. This is regardless of the fact that the males are bigger, stronger, and increasingly aggressive. Moms, sisters, and aunties acts as a group in pursuing other external females. They additionally choose their shared mate among various potential suitors wandering all through their region.

The male that is picked as a rule is one that does not act aggressively towards them and is happy to collaborate with them in defending their region. The association with a specific male might be for small time. The centre of the society is the group of related females. This is far from normal male domination.

One-male-a few female groups may take an alternate structure when predator pressure is an issue. In open fields, hamadryas baboon groups are a lot bigger, normally comprising of number of polygynous families. In such numerous one-male-and several female group societies, males are the predominant, controlling members. The adult males not only herd their own sexually developed females, but also maintain the order and protect the society from predators. This isn't abnormal for the customary Arab polygynous marriage design in which rich men get groups of females.

The gorillas have to rarely worry about predator dangers. In this manner, their communities generally comprise of a single dominant male, his mates, and their kids. As soon as the children reach the sexual maturity, they are chased off by the predominant silverback male. These males at last structure their own one-male-and several female groups. As females arrive at sexual development, they likewise leave their natal families and leave the groups.

5. Multimale-Multifemale Group

The multimale-multifemale groups are the most normal social group pattern among semi-terrestrial primates. With this example, there are no steady hetero bonds i.e. both the males and females have various mates. This is feature of savanna primates, macaques, and also of some colobus and New World monkey species.

The multimale-multifemale groups usually have a dominance order between both males and females. Every individual is positioned with respect to all other society members of the similar gender. This will in general decrease serious violence inside the community since everybody knows ahead of time who they should concede to and who must be accommodating to them. Between rhesus macaques, one's place in the dominance progression is controlled by the position of their mother.

The top positioning primates are defined by primatologists as the alpha male and the alpha female. All other community members concede to them. A female's position in the order remains with her all through life. Despite this fact, the majority of young grown up adult rhesus macaques leave their family society and at last and go along with others to discover mates. They start at the base of the male dominance hierarchy again when they leave their families to start their families. The alpha males' monkeys ordinarily mate more regularly than others.

This makes the social association externally seem as though it is one-male-a few female groups. Though, younger females frequently go to mate with males that are lower on the dominance position. The stable base of rhesus macaque societies is the groups of female relatives. They remain inside their natal network all through life and work as a group to safeguard it against external females.

6. Fission-Fusion Society

A fission-fusion society is one in which the social groups size and arrangement change all through the year with various exercises and circumstances. In chimpanzees this is the normal social pattern. The males and females enter and leave communities now and again. The males that are adults, once in a while stray and search, either alone or join a couple of different males in a hunting party. The females calmly change group from one group to the next. Thus, foraging and sleeping groups change all the time. Male chimps are the moderately steady base of the community as they in exceptional circumstances join different troops.

What permits the most part free relationships between chimpanzee communities is that they clearly identify a more extensive scope of social bonds than do monkeys? They regularly have relatives and companions in a few diverse neighboring troops. At the time as chimpanzee communities meet up, they typically exchange cordial welcome as opposed to demonstrating animosity.

Though, it would be a misstep to infer from this that chimpanzee society is always peaceful. The adult males inside every community are as often as possible occupied with complex political exercises including scheming and physical intimidation so as to climb the dominance hierarchy. They grow short term relationships with other males by mutual support, sharing meat, and allogrooming (grooming others).

The Chimpanzees are not the only primates that change groups association every now and then. For example, most adult rhesus macaque monkeys forever leave the community of their birth and to go along with others to discover mates. This is not simple since they are not warmly welcomed in their adoptive troop. Obviously, none of these monkey species change the group composition with the ease and return of chimpanzee females. Therefore, their societies are not generally described as the fission-fusion types.

7. Kinship

The relationships that are established between individuals based on family ties are described as kinship. In human societies those ties are made by marriage and shared descent from descendants. Between non-human primates, they are because of lineage. Most non-human primates clearly just perceive matrilineal ancestry. That is, they recognize who their mother is yet not their father. Socially perceived paternity is irrelevant or non-existent for them. The strongest social relationship is between mother and her little children. Chimpanzees keep up their bonds with their mother till they become adult. When they are undermined by others, even enormous male chimpanzees may go to their matured mother to be calmed by preparing.

8. Gender Differences

Behavioral variation of male and female primates comprises more than just mating behavior. The males are generally bigger and physically prevailing over females. The special feature of the semi-terrestrial monkeys and the great apes is the extreme sexual dimorphism. The way in which male and female children mingle frequently differs. After the infant age, age-based play groups are normally the basic mingling group for guys.

These adolescent male partners frequently invest quite a bit of their time in the edges of the troop areas in dynamic physical play. For instance, almost half of the waking hours of young male gorillas are spent in rough playing. This is same as to the amount of time human kid's play whenever given the chance. For most non-human primates, play is basic for creating social bonds and social aptitudes.

The young female monkeys and apes frequently socialize in more secure zones of the community territory and ae protected by their mothers and other older female relatives. Their play is normally never unpleasant and concentrates more on creating parenting abilities and social ties among the females with whom they ordinarily spend their lives. The male primates as soon as they grow up generally leave the community to join roaming bachelor groups and in the long run join different communities. There are some exceptions in the case of chimpanzees. The male chimpanzees regularly roam, the females generally change communities.

5.5. PRIMATES A NATURAL HERITAGE OF CONFLICT RESOLUTION

In the discussions on human social evolution, the origins of violence became a central theme with the early provocative description of Australopithecus as a lustful killer. The most common situations get rejected by the popular authors as per whom inborn aggressiveness, combined with male bonding in hunting and warfare, explains the human example of overcoming adversity. The extraordinary appeal of this killer ape mythology stories has been accredited to the horrors of World War two. After the war the trust in human instinct was at a low, and the view that we are dangerous insane people or a mentally unbalanced predator, putting in danger a generally amicable natural realm went down surprisingly effectively with scientist and the general population alike.

If we ignore this bigger evolutionary discussion and focus on the original research, clearly aggressive behaviour was studied as an individual as opposed to a social phenomenon. For instance, Lorenz present his controversial drive concept according to which aggressive energy develops endogenously, after which it looks for an outlet, regardless of in sports or warfare. He likewise puts focus on genetic determinants, proposing an aggressive instinct.

Figure 5.6. Bonobos Playing with each other.

Sources: Image by Flickr

The psychologists, interestingly, built up their frustration-aggression hypothesis and researched the impacts of role models and the authorities. As per ethologists, its main work was to create dispersal, a view derived from territorial fish and birds in which showing of threat indeed served to keep intruders at distance as much as possible. The psychologists too, possibly observed negative outcomes when a mouse was set in a rat cage to provoke an assault, when pain induced aggression was impelled among rodents on an electric grid, or when human subjects were told to convey high-voltage shocks to strangers.

The attention is given on hostility among people that did not know one another, students of both human and animal behaviour in this manner laid the basis for what might be known as the individual model. Because of the fact that the individual model is unaware to social setting, it fails to answer how families or societies tackle the troublesome outcomes of conflict.

The model discloses to us how animosity begins, yet not how it ends or is kept in control. In reality, in any case, by far most of hostility includes familiar people, that implies that aggressors and exploited people share a past and can be relied upon to share a future. An alternate model was required, in this way, one that sees people as socially embedded.

The primatologists were the first to advance toward this more combined paradigm encouraged by gregarious study objects. The cooperation among primates is the basis of primate societies. A few primary categories, for example, chimpanzees (Pan Troglodytes) and humans display collective intercommunity violence. Generally, the coalitions are formed inside the group with at least two individuals coming together to defeat a third. Subsequently, high-positioning people are not really the strongest, however they are the ones that can gather most support.

The universal primate task of grooming serves a significant purpose in this political field by cultivating important associations. All individuals from a group are effectively setting up and keeping up accounts of communication, which is known as social relationships. Considering monkeys and gorillas in cohesive groups, in both cage and the field, primatologists increasingly form relationships, instead of individuals, the unit of assessment.

Though these thoughts developed, a basic perception changed the manner in which we look at the social effect of conflict. The earlier research on nonhuman primates had emphasized appeasement and reassurance gestures and had alluded to relationship fix aftereffect of an episode after battles. The latter situation was named and observationally characterized as the result of the incident on the planet's biggest chimpanzee colony at the Arnhem Zoo, in the Netherlands. At the point when the alpha male savagely assaulted a female, different primates went to her guard, causing continuous shouting and chasing in the group. After the chimpanzees had calmed down, a tense silence followed and was broken when the entire colony burst out hooting.

Amidst this mayhem, two chimpanzees kissed with their arms folded over one another. These two male and female chimpanzees that have taken part in the fight in the past are the same in this fight. There was an aggressive standoff, but then the compromise was reached, and it is displayed as a friendly reunion between previous opponents, the data on several occasions demonstrated the pattern to be continuous, normal pattern pf the social life of chimpanzees in the Arnhem chimpanzee zoo. In the decade of 1970 there were other developments, this meant a strong framework for the investigation of conflict resolution had come into place spinning around the related three components

- signs of a calming function of grooming and other body contacts
- acknowledgment of long-term social connections and their survival worth
- The display of an association between aggressive conflict and resultant inter-opponent reunions, defined as reconciliations.

5.6. PRIMATE ABILITIES IN THE SOCIAL REALM

5.6.1. Knowledge of Other Individuals

The researchers empowered by playback studies to simulate signals from particular people or connections between particular partners utilizing the objective creature's response as a list of what it knows. For instance, playback studies have been utilized to think about individual acknowledgment in different animal species. Ground squirrels (Spermophilus richardsonii), marmots (Marmota flaviventris), and meerkats (Suricata suricatta) have all been appeared to recognize the personality of others from vocalizations.

For marmots and ground squirrels, this enables people to decide if alert calls were given by reliable or unreliable group members according to Blumstein et al., 2001, Hare and Atkins, 2001, however the capacity of this segregation capacity isn't clear for meerkats according to Schibler and Manser in 2007.

The monkeys know about the individual identity of others as demonstrated by the playback experiments, and not just those people inside their very own group. Female vervet monkeys (Chlorocebus aethiops) stay in their natal groups all through life. At the time when calls for females from groups who had been neighbors for their entire lives were broadcasted, the aimed people responded more strongly when the playback experiment recommended the caller has moved out of group (Cheney and Seyfarth, 1982).

The sound of a crying child is heard as soon as by Vervet monkey, they are not just increasingly responsive toward their own baby's cry, yet when they are played the calls of different babies they look toward the genuine mother, demonstrating they perceive mother-baby connections (Cheney and Seyfarth, 1980). This result was confirmed in captivity for long-tailed monkeys utilizing match-to-test (Macaca fascicularis) and stretched out to the sibling relationship (Dasser, 1988), demonstrating that at least two types of kin relationship are recognized. In addition, in the wake of losing in struggle, monkeys of a few species show their animosity toward more weaker monkeys, yet explicitly toward relatives of the individual who defeated them (Cheney and Seyfarth, 1989, Judge, 1982), demonstrating more knowledge of family relationships.

Figure 5.7. Monkey mother with baby

Sources: Image by Pixabay

There are some Old-World monkey species that give explicit welcome calls to dominants and subordinates. At the time when target vervet monkeys hear playbacks, their response relies upon the rank connection among themselves and the caller. When the caller is dominant to themselves, and gives a "call to dominant," their agitated reaction suggests that they expect a high-ranking individual nearby; whereas if the same monkey gives a "call to subordinate" or if a monkey subordinate to themselves gives a "call to dominant," they give little reaction: they understand the ranks of third parties relative to each other (Cheney and Seyfarth, 1982a).

Playback can even be utilized to mimic connections between outsiders: when baboons (Papio ursinus) hear sounds that propose a rank reversal in the dominance hierarchy, they respond with astonishment, explicitly if the inversion recommends that the normally stable rank relationships between matrilineal families have changed (Bergman et al., 2003). The baboons clearly "listen in" on social relationships when playback reproduced an associating pair moving separated, which would ordinarily signal a mating opportunity, different males rushed to set out towards the female (Crockford et al., 2007).

5.6.2. Social Tactics

Just as great proof that primates perceive people, family relationship, rank, and the outsider connections between others, there is currently broad proof that numerous types of primate normally send subtle or manipulative social strategies during intragroup challenge with these people.

The primate and also for other social animals the dominance rank is very important, rank is more averse to be a component of physical power, frequently deriving rather from the support of third-party relationships between others (and this isn't special to primates: e.g., zebras [Schilder, 1990]). In macaques, the female monkeys may support her child against others positioning below her, and specifically her youngest child, the outcome is that child reach dominance ranks just beneath that of their mom, with the youngest the highest positioned (Sade, 1967).

The grooming is utilized to develop alliances that collect future advantages in different currencies in all old-world monkeys (Dunbar, 1991), fluctuating the exchange rate in a biological market (Fruteau et al., 2009). The social bonds that are formed are of long term pay in reproductive terms: the child of baboons that forms solid social bonds with different females live more, independently of health differences between the guardians, and the impacts endure into children adulthood (Silk et al., 2009).

Monkeys are very choosy with whom they structure unions with, focusing on their grooming and other affiliative collaborations explicitly at the most valuable partners, for example, the individuals who are dominant or who offer different advantages (Cheney, 1978, Harcourt, 1992). A similar particularity in allotment of grooming to the most valuable partners has as of late been found in meerkats (Kutsukake and Clutton-Brock, 2010) and is maybe likely in every social species that show allogrooming. The more responsive are the Vervet monkeys to the (experimentally played back) vocal signals of individuals who have as of late prepared them, if they are not close family, to whom their responsiveness is very high (Seyfarth and Cheney, 1984).

At the time when their most significant coalitions are upset by rivalry, primates and monkeys of different species make efforts towards compromising, they are bound to start grooming with an recent opponent than anticipated from control perceptions (Cheney and Seyfarth, 1989, Cords, 1992, de Waal and van Roosmalen, 1979, de Waal and Aureli, 1996, Kappeler, 1993).

The vocalizations can be utilized successfully in baboons to accommodate with previous adversaries, at a distance (Cheney et al., 1995); and even a friendly sound given by the nearby family of an ongoing rival serves to accommodate recent opponents (Wittig et al., 2007b). The support in aggressive contest can successfully be given by vocalization in the same manner, making a vocal partnership (Wittig et al., 2007a), and baboons can gather the objective of a vocalization from the specific situation (Engh et al., 2006). Supporting and fixing partnerships among non-family proposes that primates somehow or another comprehend that grooming, maybe as a genuine pointer of time contributed (Dunbar, 1992b), can be exchanged in the desire for future advantages a type of complementary unselfishness (Trivers, 1971).

One of the most interesting manners by which primates control their allies is appeared in "policing," found in some macaque species, where an influential individual mediates to separate battles among subordinates (de Waal, 1989): the expense of policing falls just on the intervener, however the advantages collect to all, making policing generally unlikely to advance (Flack et al., 2005). Flack and her teammates have utilized the method of knocking out specific individuals in a huge group of pig tail macaques (Macaca nemestrina) to examine experimentally in the case of policing is successful, finding that rare policing by a couple of influential people fundamentally protects the stability of social communities despite irritations (Flack et al., 2006). Regardless of whether this type of niche construction (Odling-Smee et al., 2003) depends on innate inclinations in pig tailed macaques or some comprehension of the social mechanisms is obscure.

A solid case for more than this has been put forth in the defense of Taï chimpanzees of Ivory Coast, a population known for large amounts of hunting; coordinated and distinctive individual roles have been depicted, containing driver and ambusher (Boesch, 1994). Though at Ngogo, Uganda, the study site where hunting is considerably more important and comparably seems helpful, the researchers were not able to ensure that more than individual selfish tactics were included (Watts and Mitani, 2002). In numerous social carnivores the cooperative hunting is common phenomenon (e.g., spotted hyaenas [Crocuta crocuta], a few population of lions [Panthera leo], and wolves [Canis lupus] [Grinnell et al., 1995, Holekamp, 2006, Mech, 1970, Stander, 1992]) and at minimum one birds, the Harris' hawk (Parabuteo unicinctus; Bednarz, 1988). In spite of the fact that Stander portrays female lions assessing the developments of others during chases, the degree to which participation depends on any comprehension of the methodologies of others is generally not known. A rich special case is crafted by Drea and Carter (2009) who brought out sets of spotted hyaenas with a rope-pulling task that must be tackled by collaboration. They found that hyaenas participated rapidly and more than once yet in addition improved after some time as individuals progressively assessed the activities of the other and demonstrated settlement toward the naive accomplice.

5.6.3. Perceiving Others

The animals of different varieties are recognized to differentiate a series of conspecific shows and body postures, consisting of facial expressions; the increased the degree of manipulation of facial musculature, the larger the quantity of facial expressions that may also want to be recognized. Though, shows and expressions are to a massive extent beneath the control of the signaler, and what facts it desires to share; it would be of great utility to a competitor, therefore, if the involuntary or inevitable moves of the eyes, head, and body that exhibit the gaze path and (presumably) the focal point of an individual's visual interest could also be read. Accordingly, a good deal effort has been committed to learning primate capabilities at gaze following and reading the attention of others.

Figure 5.8. Monkey mom teaching baby to climb the Rope.

Sources: Image by Wikimedia

The gaze following also shows specific places, objects, or things to do that are the centre of attention of others. Oddly, it has been distinctly challenging to show that species obtain beneficial information of this type from gaze following for instance, in finding food in an object-choice experiment, domestic dogs far outperform chimpanzees (Bräuer et al., 2006; Krachun and Call, 2009). According to Hauser in 2007 even monkeys, able to use human pointing and different communicative cues, and to observe gaze, do no longer seem to be capable to use gaze as a trick. The object selection on the basis of gaze in present times following has been shown in lemurs, Eulemur spp. (Ruiz et al., 2009), even though no longer at high levels of success.

The learning of the perceptual importance of some other individual's point of view has been much more readily shown in gestural communication. The Great apes are very attentive to the ability of their audience to discover gestures, selecting tactile or auditory modality when the target audience is not paying attention, and silent visible gestures only when they are (chimpanzees [Tomasello and Call, 2007, Tomasello et al., 1994], bonobos [Pika, 2007a], gorillas, Gorilla gorilla[Genty et al., 2009, Pika, 2007b], orangutans, Pongo spp. [Liebal, 2007, Liebal et al., 2006]). Furthermore, some individual apes have proven that they understand the utility of hiding embarrasing revealing signals, such as a chimpanzee who was once noticed to disguise his penile erection from a dominant male until it subsided (de Waal, 1982), or a gorilla who typically hid her play face in order to shock a play associate (Tanner and Byrne, 1993).

5.6.4. Predicting and Understanding the Actions of Others

The capacity to predict what others are likely to do next can be indispensable for survival. It is evident that a huge range of animals are properly in a position to extrapolate the present-day behaviour of conspecifics, predators or prey, to expect what their next movements may be; however, the evidence that they may also do this on the basis of appreciation (representing unobservable causal factors) is more tough to obtain.

In the situations of social cognition, the primary difference for humans is between unintentional and intentional action, and whether or not nonhuman primates experience in a similar way has been investigated in a numerous way. The captive chimpanzee, when a person accidentally spilled its prized orange drink and gave it nothing, in consequence continued to beg from that person; however, if the loss regarded deliberate, with the juice simply poured away, the chimpanzee avoided that person in the future according to Povinelli in 1991.

REFERENCES

1. Byrne, R. and Bates, L. (2010). Primate Social Cognition: Uniquely Primate, Uniquely Social, or Just Unique?. *Neuron*, 65(6), pp.815-830.
2. https://pdfs.semanticscholar.org/30fd/088e919002dddb1be53f824ab527a8da45a0.pdf [Accessed 27 Aug. 2019].
3. https://www.nature.com/scitable/knowledge/library/primates-in-communities-the-ecology-of-competitive-59119961/ [Accessed 27 Aug. 2019].
4. Matsuzawa, T., Hockings, K., Humle, T. and Carvalho, S. (2012). Chimpanzee interactions with nonhuman species in an anthropogenic habitat. *Behaviour*, 149(3-4), pp.299-324.
5. https://www2.palomar.edu/anthro/behavior/behave_2.htm [Accessed 27 Aug. 2019].

Chapter 6 Conservation of Primates

"Without collaborative and immediate action, it is possible that our great-grandchildren will never see the beautiful ecosystems and wildlife that many of us are fortunate to live in and research."

-Christopher Golden

CONTENTS

The topic Conservation of Primates discusses the basic concept of primate conservation and its habitats, the different reasons of arising needs to conserve the primates like habitat loss, disease, and hunting. It includes role of primates in our environment and different reasons to conserve the primates. The readers will get to know about the various conservation efforts that were made in this direction by different countries at the global level under one platform and that led to formation of IUCN. The different primate threats like issues of poaching, illegal wildlife businesses and the threat to both humans and primates of communicable diseases.

This chapter also throws light on the advantages and importance of saving primates as primates and human beings are both dependent on each other and primates also saves our environment.

6.1. PRIMATE CONSERVATION

The world's population of primate species in which more than two third are in danger of extinction. In the upcoming 66 years there is a chance that more than 66 percent of all primate species will become extinct. The different activities by humans are putting in danger the various nonhuman primates through habitat loss. The fight for mistreatment of the resources restricts the quantity of people with the declining habitats which in a better way survive and thrive, building the consequence of decline within the system. The loss of lives is caused by loss of habitats; with the loss of food and shelter, and together with animals.

In order to provide the essential activities such as the enriching and fertilizing the soil through the time period of their lives (seed spread through their waste matter) in the absence of the animals are disrupted. Every environment contains precisely the correct quantity, combination, and variety of plants and animals to make a beautiful, independent relationship.

The biodiversity exhibits thanks to reciprocity of species on one another, the world and therefore the creatures live and thrive due to diversity.

The activities related to agriculture and urban development are taking place and for this purpose the rainforests are being cut down, the earth is losing the most valuable resources, the trees that provides oxygen. The rainforests are said to be earth's lungs. The air quality gets affected globally by the decrease of rainforest. As the rainforests are reducing in size, subsequently the fewer trees, less oxygen is produced, putting all the species dependent on oxygen including the humans are at peril.

Figure 6.1. Primate in eastern European Forest.

Sources: Image by Flickr

The huge quantity of carbon diffused into the atmosphere by human activity impacts more the global weather. These bad effects are happening very fast. These damaging issues are because of human activity, they are avoidable, repairable, and can be reversed and this is good news. The time is passing, and human beings must take steps now to avert more damage. The human beings cannot wait and look for someone other to do something. The human being shall become the change that everyone is being expected.

What are we, without the, other animal species, that includes primates or otherwise. The entire human race amongst several of the earth citizens, a component of the planet interdependence on each other. The plant and animal that is every species has its part in maintaining ecological balance. There are some species that will become extinct if there is no change in the behavior of human beings about their benefit to human beings and the effects of the daily selections on the primates.

The monkeys found in Africa are Old World monkeys. Many of them live all over the tropical and equatorial zones. There are very less monkeys that live in the semi-arid and rocky northern reaches of Africa. The distribution of some species spans into southern Africa. Some monkeys live in the human colonies there are some monkeys that lives in cities.

In Africa's nonhuman primate species, some live only on trees, some along river and mangrove banks, and few live on the ground. There are species that are mostly active in daytime, but some gets active in night-time.

The access to resources gets distributed and competition gets reduced by living at different heights, whether in trees or on the ground, and separate activities between day and nighttime. This is one of the behavior that nature makes sure that everyone has access to all of the gifts of the Earth.

The Asia monkeys are old world monkeys. They are mainly found in southern Asia, with some species as far north as Japan and northern China. The Asian nonhuman primates mostly live in tropical habitats. In the human settlements some monkeys live in cities. There are some that live in the far-off snowy mountains of China and Japan.

Figure 6.2. Monkeys found in Asian Region.

Sources: Image by Maxpixel

In Asia some monkeys live only on trees, some along river and mangrove banks, and few live on the ground. There are species that are mostly active in daytime, but some gets active in night-time.

The monkeys found in Latin America are called as New World monkeys. In the regions of southern Mexico to central South America they live. The nonhuman primate species of Latin America reside in tropical habitats, mainly rainforests, in the equatorial zone. Few live-in cloud forests. There are few that lives in human settlements that include cities.

6.2. PRIMATE CONSERVATION: IS THE CUP HALF EMPTY OR HALF FULL?

There are around 600 types and subtypes of primates living at the present time and on an average half of them are in hazard of extinction; an approximation that has increased evoking serious concern over the preceding decade according to Mittermeier in the year of 2009.

In fact, one sub variety in West Africa, Miss Waldron's red colobus (Procolobus badius waldronae), may also be extinct according to Oates in the year 2000. Moreover, the range of identified serious threats to primate survival has grown over the last decade.

A decade ago, whilst ailment was once identified as an essential region of study, it was not thought as a thing that ought to threaten primate populations with nearby extinction. In difference, these days there are a number of nicely documented instances of dramatic primate population decrease induced through disease. For example, Bermejo in 2006 and it is thought to have an imperative aspect in conservation planning, especially for those populations that are in risk of contracting ailments from exposure to humans. For example, ape ecotourism. In the same manner, in the last decade, climate change has long past from something largely not noted by many countries to an event of grave concern.

6.2.1. Habitat Loss

The Food and Agriculture Organization of the United Nations (FAO) in the year of 2010 has approximated that 16 million hectare per year of forest was lost around the world in the decade of 1990; on an estimation 12.5 million hectare per year was lost in countries with primate populations, the region larger than the state of Mississippi, or simply smaller than Greece according to Chapman and Peres in 2001.

According to world level estimation of primate densities this would have resulted in the loss of 32 million primates per year as per Chapman and Peres in the year of 2001.

In difference, in the last decade, the degree of deforestation has diminished globally to approximately thirteen million hectare per year and reforestation and natural increase of forests in some countries notably decreased the total loss of forests according to FAO in 2010.

The total loss in forest region from 2000 to 2010 is approximated at 5.2 million hectare per year, down from 8.3 million hectare per year in the last decade. Some of the increased degree of reforestation can be credited to decreasing human fertility rates and a growing urbanization fashion that is happening in many developing nations according to Jacob in 2008, Wright and Muller-Landau in 2006.

These elements have resulted in the reduced demand and even abandonment of agricultural lands in many areas of the world. The variation in human demographic patterns and forest cover differs dramatically depending on which area of the world one is thinking about, and these parameters provide each a capability of predicting future forest loss or gain and are a necessary factor in conservation planning at a local level.

Figure 6.3. Monkey primate Mother and Baby.

Sources: Image by Maxpixel

According to Bauche in 2007, Jacob in 2008 the interviews and assessment of demographic information of the residents discloses that the lack of high schools and job opportunities in the village regions influenced migration of the youth to the closest city of Puerto Vallarta; a tourist center with higher economic opportunities. This migration left areas un-used and tropical forests regenerated.

Though decreasing indications of habitat loss are an encouraging signal for primate conservation, deforestation is also taking place at an alarming rate, and robust efforts need to be made to encourage forest regions conservation and regeneration.

It is stated that both because of the direct effect of any habitat loss on primate populations and because of how it indirectly affects primates by meeting with other drivers of populations change. For example, deforestation and other land use modifications are estimated to have released 150 billion metric tonnes of carbon into the environment from the 1850 that is approximately one fifth of the total carbon in the surroundings according to Houghton in 2003.

The deforestation presently accounts for about 18 percent of world carbon emissions, which is the third biggest source of emissions (IPCC 2007, Keith et al. 2009), and it is this carbon dioxide that is one of the main causes of climate changes. The climate change is estimated to have strong negative consequences on some primate species.

6.2.2. Disease

The surprising appearance of ailments like swine flu in humans and wildlife, the devastating impacts that ailments like Ebola have had on both human and wildlife communities, and the significant social and economic cost created by viruses like HIV, undermines the need to understand the ecology of infectious diseases.

It is known that monkeys and apes generally share diseases with humans because of their shared evolutionary history, learning the ecology of infectious diseases in non-human primates is of supreme importance for both primate conservation and public health.

It has been estimated by outbreaks of the Ebola virus that trace their birth to zoonotic transmissions from local apes according to Leroy in 2004. The most dramatic illustration of the possible effect of ailment on primates is the recorded decrease in endangered ape populations. For example, in the areas of the Congo Basin there have been some reports that state that the decrease in the population of gorilla and chimpanzee, with recorded decrease of affected populations being as high as 80 to 99 percent according to Bermejo in 2006, Huijbregts in 2003.

In the decade of 1980, the overriding viewpoint on the remedy and prevention of infectious diseases was one of optimism according to Cohen in 2000, and immunization and antibiotics had been viewed adequate for combating infectious diseases. In fact, the eradication of smallpox in 1979 has created the hope that, through collective world action, most principal diseases could be eradicated as per Fenner in 1983.

This optimism was broken by the growing occurrence of antibiotic resistant bacteria and the rise of diseases such as Ebola, HIV, multi-drug-resistant tuberculosis, malaria, and enterohemorrhagic E. coli, and much difficulty in eliminating diseases globally despite big efforts. The diseases are now seen as a serious risk to primate and human populations, and a thought of how environmental change might encourage contact between people and non-human primates, and make bigger the possibility of sharing infectious diseases, is now very important to both conservation and human health planning according to Bonnell in 2010.

Moreover, the awareness has been made that disease outbreaks can be promoted by way of anthropogenic modifications to the environment, such as logging according to Gillespie in 2005 and climate change (Chapman et al. 2010), that is reason for increased concern.

6.2.3. Bushmeat Hunting

In all possibility the hunting been a principal driver of primate population decrease for many decades, it was not till the early years of the decade of 1990 that teachers and conservation biologists began to comprehend how devastating an impact was having on the neighborhood abundance of some primate species. For example, a market survey in two cities in Equatorial Guinea, West Africa, with a mixed human populace dimension of 107,000, recorded 4,222 primate dead bodies on sale over 424 days (Fa et al. 1995).

According to Peres in the year of 1990, he recorded that a single family of rubber tappers in a far off forest area of the Brazilian Amazon killed more than 200 woolly monkeys (Lagothrix lagotricha), 100 spider monkeys (Ateles paniscus), and eighty howler monkeys (Alouatta seniculus) over the time period of 18 months. There is an approximation that the yearly wild meat harvest is 23,500 tones in Sarawak, Malaysia, 67 to 164 thousand tonnes in the Brazilian Amazon, and 1 to 3.4 million tonnes in Central Africa according to Milner-Gullard in the year of 2003.

The estimate from Central Africa is nearly six times the most sustainable rate of harvest (Milner-Gullard et al. 2003). The trade and consumption of bush meat is generally incorrectly thought as being a local problem, with the meat being consumed solely by human beings in the region. The bush meat business is a large global business. For example, it is estimated that about five tons of bush meat is smuggled in personal baggage through Paris Charles de Gaulle airport every week (Chaber et al. 2010). This weight in bush meat is equal to shifting approximately 570 cows via this airport a year.

Subsequently the bush meat business is now not simply a local problem, there are excellent possibilities to coordinate advances in both development and conservation. When bush meat is not eaten in local areas it needs transportation that generally takes the shape of loading carcasses onto logging trucks. This offers the unique probability for governments and conservation groups to improve strategies when road systems are being developed or logging concessions assigned, which decreases destructive results on forest flora and fauna according to Wilkie in 2000.

6.2.4. Climate Change

The earth weather has warmed by 0.6 degree Celsius over the last 100 years, and some prediction suggests that the climate could heat by 6 degree Celsius this century (IPPC 2001, Walther et al. 2002). There is a great deal of apprehension about plausible impacts of climate change on primate populations and their tropical environments, however genuine data that can be bought to bear on the effects of climate change are very limited.

Though, Dunbar in 1998 has estimated how gelada baboons (Theropithecus gelada) would respond to climate warming and presented that the lower altitudinal limit of the species would grow by approximately 500 m for every 2 degree Celsius increase in world mean temperature; as an end result the land regions available will become half and largely fragmented. The 7°C upward growth in temperature would be sufficient to decrease the species' population size from its current size of approximately 250,000 to about 5,000 geladas on a few isolated mountain peaks the place their long-term survival would be unlikely.

Also, the direct effects of climate change on primate populations, primates may additionally be affected by both the loss of unique plant species or changes in the leafing, flowering, and fruiting patterns of plant communities. With the help of a sequence of computer modelling exercises, McClean in 2005 studied the impact of estimated climate trade on the distribution of 5,197 African plant species. For 81 to 97 percent of these species, areas of suitable climate are projected to decrease in dimension and or shift in location, many to higher altitudes, and 25 to 41 percent of species will lose all their regions by 2085.

The models suggest dramatic modifications in the forest that stretched from Guinea in West Africa all the way to the Rwenzori Mountains on the east side of the Democratic Republic of Congo that presently covers some of the most extensive blocks of leftover forest in Africa.

The variation in the climate will also have interaction with other drivers deciding primate population sizes. Human scientific experts have lately come to be concerned as to whether world warming will cause increased rate of infectious diseases and with their wealth of clinical data, are properly ahead of primate ecologists at documenting trends.

For example, connections between climate and disease are properly established, and many ailments happen during certain seasons or erupt in relation with unseasonable conditions. For example, in sub-Saharan Africa, meningococcal meningitis epidemics erupt throughout the warm dry season and subside soon after the start of the rains (Patz et al. 1996).

In the United States, 68 percent of waterborne disorder outbreaks had been come first by rain activities above the 80 percent percentile as per Hunter in 2003. The similar connection between the environment and diseases possibly present for primates, signifying climate change might alter ailments affect, primate populations as well.

It is a fact that many primates are endangered by human activities and that tremendous efforts need to be positioned in conserving the forests they inhabit. However, we trust that there is purpose to be constructive as the rate of forest loss is decreasing, an excellent deal is now regarded about primate diseases, and programmes to minimize the bush meat industry are being efficiently implemented (Wilkie et al. 1992).

This optimism should not lead us to decrease our efforts; instead this knowledge of the threats offers great possibilities for scientists to make widespread contributions to primate conservation. In many respects, the research conducted over the last decades has established very clear goals that can be met to promote primate conservation; for example, by providing individuals sustainable substitutes to the bush meat trade, decreasing deforestation and promoting reforestation, growing parks, and making rules for lowering the threat of disease transmission between humans and primates, and decreasing carbon emissions to curb climate change.

6.3. ROLE OF PRIMATES

The primates are the closest living species to human beings that have been used as leading species by conservationists to bring focus, public attention to the latest large-scale destruction of tropical forests. The saving of the natural habitats of primates, the human beings can save lots of species of less charismatic, though equally endangered, flowers and animals in the tropics.

The Primates have also been focused as top species because they offer many ecosystem services, such as seed dispersal and pollination. A full complement of dispersers and pollinators are fundamental for tropical forests to regularly reproduce and sustain diversity, and hence maintain ecosystem health for the plants, animals, and people that depend upon them.

There are seven countries that have been focused as primate conservation priorities due to the fact of their large primate diversity and risk of extinction: The Democratic Republic of Congo (17.6% threatened), Peru (27%), Brazil (36.4%), Madagascar (40-42%), Nigeria (42.3%), Indonesia (84.1%), and Vietnam (86.4%) as per IUCN in 2008.

6.4. THE VARIOUS CONSERVATION POLICIES

In the Sixties, the IUCN that is International Union for the Conservation of Nature in the decade of 1960 has created Red List of Threatened and Endangered Species as a conservation tool to recognize world extinction hazard, at the same time providing species-specific data on distribution, ecology, and threats to persistence. At minimum 49 percent of all primates are recognized as vulnerable to extinction, a number greater than any different mammalian order (Mittermeier et al., 2007; IUCN Red List, 2008).

The world countries gathered in Brazil in 1993 to talk about the disaster of biodiversity loss and outlined a design of goals and strategies to help slow the losses. The Primatologists, with a long-term presence in the tropics, used this blueprint as a guideline and started a large-scale conservation effort in primate habitat countries. A series of primate conversation plans for particular geographical areas had been created by the IUCN.

Figure 6.4. Lion Tailed Macaque Western Ghats.

Sources: Image by Pixabay

The Integrated Conservation and Development Projects were funded in particular regions in Africa, South America, and Madagascar. The policy that was developed was country specific, such as the Durban Vision in primate-rich Madagascar, in which the President promised to triple the covered area in six years. Because of the extra increased interest that climate change has gathered, the international meeting took place in Bali in December 2007.

This meeting resulted in the creation of the United Nations Collaborative Programme on Reducing Emissions from Deforestation and Forest Degradation in Developing Countries (UN-REDD Program), which hopes to limit environmental degradation through funding initiatives to stop tropical forest area burning and start habitat restoration. In order to increase public consciousness of the terrible condition of some of our closest relatives, taxon-specific initiatives were also launched, like the 25 most severely endangered Primates list, which used to be announced to the media in 2000 and every other year since, and the UN Year of the Gorilla in 2009.

All these efforts have helped to carry the urgency of retaining primates and their habitats to the attention of governments and the public before it is too late. They have also modified the nature of primate conservation efforts. The Primatologists have lately started to hire new conservation tools, which are proving fundamental for effective, measurable, speedy results.

6.5. WHY CONSERVE PRIMATES?

6.5.1. Primates Encourages Human Health

The Primates from long time have been considered important to study for the betterment of human health. Though primates contain a small percentage of animals used in biomedical work, their very close genetic and physiological similarity to humans makes them especially valuable in developing treatments for and vaccines towards human ailments (Bontrop 2001; Carlsson et al. 2004; Sibal and Samson 2001). Certainly, researchers studying a vast variety of ailments and disorders think about primates to be irreplaceable to research that is finally aimed at bettering human health and health (Bennett 2015; Capitanio and Emborg 2008; Evans and Silvestri 2013; Joyner et al. 2015; VandeBerg and Zola 2005).

Figure 6.5. Eastern lowland gorilla in Congo.

Sources: Image by Wikimedia

The number of primates used in biomedical research is raised in captivity for this cause (California Biomedical Research Association 2015). However, wild populations are sometimes used as source populations in exceptional instances the place where the captive-bred primates are inappropriate. (For example, Home Office 2004; United States Department of Agriculture 2013). The growing danger from increasing infectious diseases and the fastly altering environmental conditions resulting from world weather change might increase the significance of wild primate populations as sources of research subjects.

For example, primate populations that has natural immunity to novel pathogens may also offer special insights that support combat future human diseases. The extinction of wild primate populations should mean the loss of information indispensable to human survival in a future of rising infectious diseases and world climate change.

6.5.2. Primates Provide Benefits to Local Communities

The human beings living nearby to wild primate populations can gain vital benefits from them. The primates and other sources of wild meat can act as necessary food resources for communities living inside or adjacent to tropical forests in some regions (Brashares et al. 2011; Millner-Gulland et al. 2003).

For example, feasting on meat from wild animals, together with primates, used to be related with significantly decreased occurrence of Anaemia in young people residing in villages around the Makira Protected Area in northeastern Madagascar (Golden et al. 2011). These types of hunting are though, usually unsustainable (Cowlishaw and Dunbar 2000; Fa et al. 2002, 2005; Fa and Tagg; Golden 2009) and can cause the local extinction of species according to Nunez-Iturri in the year of 2008.

However, sincerely sustainable administration of primate populations for food would, by definition, make sure their long-term persistence and therefore could conceivably be used as a justification for primate conservation beneath certain, distinct conditions (Cowlishaw and Dunbar 2000; Crockett et al. 1996; de This et al. 2009; Ramirez 1984).

This disagreement is off course incompatible with different justifications for their safety, stressing the complexities inherent to most conservation and reminding us that the groups that share a common purpose may additionally do so for the diverse reasons.

In some areas, conservation of a unique primate population might give economic benefits to local communities (Siex and Struhsaker 1999; Davenport et al. 2002). For example, large income is generated in some communities from primate and rainforest tourism (Adams and Infeld 2003; Archabald and Naughton-Treves 2001; Kirby et al. 2010); such tourism might also in turn encourage conservation in some specific situations (Pusey et al. 2008; Kirby et al. 2010; Savage et al. 2010).

In addition to tangible benefits they may additionally provide, primates might have cultural or religious significance for humans living in the regions that are close to forest according to Fuentes and Wolfe in 2002; Riley 2010. For instance, the Hanuman langur in India is viewed holy in the Hindu religion and the Iyaelima humans in the Democratic Republic of Congo have taboos that stop them from eating bonobos according to Fuentes and Wolfe in 2002.

In such situations local humans have likely been living in close proximity to wild primates for centuries (for example Tutin and Oslisly 1995), and in some circumstances can be effective advocates for primate conservation. Therefore, for local individuals, extinction of close proximity primate populations will restrict sources of wild meat, limit financial opportunities, or decrease deeply held cultural beliefs.

6.5.3. Primates Serve Key Ecological Functions

The Primates regularly do important ecological functions in the ecosystems they inhabit. First, primates give pollination services in some ecosystems (Carthew and Goldingay 1997; Gautier-Hion and Maisels 1994; Janson et al. 1981). For example, Kress et al. (1984) has done an extensive research on the relationship between the traveler's tree (Ravenala madagascarensis) and ruffed lemurs (Varecia variegata) and determined that the system showed features of two a co-evolved plant–pollinator relationship.

Second, primates are widely recognized to be necessary seed dispersers (Chapman 1995; Lambert and Garber 1998; Norconk et al. 1998; Sato 2102; Tutin et al. 1991). In some plants, seed germination rates are positively influenced by passage through the primate gut; other plant species rely totally on primates for dispersal (Chapman and Onderdonk 1998; Wrangham et al. 1994). There is mounting proof that local extinction of primates appreciably alters plant species composition (Effom et al. 2013a; Nunez-Iturri et al. 2008; Vanthomme et al. 2010).

Third, primates are vital seed predators in some ecosystems (Peres 1991; Peters 1993; Norconk and Veres 2011), and whilst it has to date obtained little attention, it is viable that primate seed predators may also assist preserve plant species diversity by excessively preying on seeds of frequent plant taxa (cf. Paine and Beck 2007; Power et al. 1996; Terborgh 2012).

Fourth, folivory through primates can affect the mortality, fertility, and growth rates of tree species (Chapman et al. 2013). Fifth, the presence of primates can affect community structure across more than one trophic level. For example, the loss of primates because of hunting in Nigerian tropical forests has resulted in alterations of the relative abundances of other mammals, with cascading effects on plant communities (Effom et al. 2013b).

Sixth, primates are vital prey species in some ecological communities (Isbell 1994; Hart 2007); some species, most pretty chimpanzees, can additionally have vast impacts as predators on primates and other animals (Stanford 1995; Teelen 2008). Finally, primates might also play an important role in buffering in opposition to the harmful effects of global climate change.

According to Howe in the year of 1986, the Primates are generally the key dispersers of large-seeded plant species, and large-seeded tree species frequently have higher carbon densities than trees with small seeds as per Queenborough in the 2009; Wright and others in 2007. Therefore, the existence of primates encourages the appropriation of additional carbon in tropical forests, which assist as key buffers towards global climate change according to Van der Werf and others in 2009. These examples establish that primates play an essential role in retaining well-functioning ecosystems.

It has commonly been difficult to find out whether or not primates serve keystone features in ecological systems, in part because it is uncertain to what extent the ecological roles of primates would be escaped through other taxa were primates absent (e.g. Chapman and Onderdonk 1998; Gautier-Hion et al. 1985; Poulsen et al. 2002; Russo and Chapman 2011; Chapman et al. 2013).

The increasing evidence however shows that at minimum in some systems primates serve uniquely vital roles, and that their loss has giant effects that are not offset by other taxa (Effom et al. 2013a, b; Muller-Landau 2007; NunezIturri et al. 2008).

The Primate conservation is consequently very important to sustain intact ecosystems and the many services these ecosystems offers to people, that includes clean and steady water supplies, prevention from floods and landslide, pollination, balanced micro-climates, and buffering of world warming according to Wich in 2011.

6.5.4. Primates Provide Unique Insights into Human Evolution

The human beings are basically primates and therefore the safety of wild primate populations conserves our capability to learn about the ecology, sociality, and behavior of our close relatives according to Boyd and Silk in 2012; Fleagle in 2013. The deep knowledge of human beings is not possible except setting the evolution, biology, and culture in large phylogenetic context.

The extinction of a primate species would lower the capability to apprehend ourselves, our evolution, and our place in nature. For example, thought of humans in the context of non-human primates has increased our understanding of human cognition (Matsuzawa 2001; Tomasello 2009), genetics (The Chimpanzee Sequencing and Analysis Consortium 2005; Patterson et al. 2006), communication (Savage-Rumbaugh et al. 1998; Tomasello 2008), aggression (Smuts 1992; Wrangham and Peterson 1996), reconciliation (Aureli et al. 2002; de Waal 2000), ecology (Hill 1982; Ulijaszek 2002), and a lot more.

The Studies of existing primate tool use, hunting, cultural traditions, and diet importantly inform reconstructions of human evolution (e.g. McGrew 1992; van Schaik et al. 1999; Matsuzawa 2001; Boyd and Silk 2012). Due to this most primatologists in the United States, and many in Europe and

Japan, are associated with academic departments or institutes principally devoted to the study of anthropology and human evolution.

The Studies of great apes has been of special interest, due to their close phylogenetic relatedness to human beings (de Waal 2005; Knott 2001; Semendeferi et al. 2002; Wrangham 1987; Wrangham and Pilbeam 2001), however different taxa have also been argued to provide very important learning into the evolution of human behaviour (DeVore and Washburn 1963; Kinzey 1987).

The loss of any primate species, but precisely an ape taxon, would both abate our ability to differentiate homologies (characters shared primarily based on frequent descent) from homoplasies (characters advanced independently through convergence) in hominoid evolution and restrict our understanding of the extant of variation in some traits.

Bonobos show off a number of characteristics that are in difference unambiguously with regular patterns considered in other great apes: bonobo females are more social, shape stronger bonds with one another, and have to face less dangers of sexual aggression or infanticide as compared to other apes (Hare et al. 2012; Stumpf 2006; Surbeck et al. 2012). These studies helped in starting the examination or investigation of and understanding for the importance of female social relationships as well as the social function of sexuality in apes, thereby broadening conceptions of the series of variation possible in the lineage producing chimpanzees, bonobos, and humans (Kano 1992; Parish 1994; de Waal 2005). The Protection of primates saves valuable facts about us and our past.

6.5.5. Primates Are of Immense Biological Interest and Importance

The primate order is a varied group and displays great variation in ecology, social system, and behavior (Smuts et al. 1987; Kappeler 1999; Mitani et al. 2012; Rylands and Mittermeier 2014). The Primate species span at minimum 4 orders of magnitude in body size, devour a vast range of diets, shows the most varied group of locomotor adaptations of any animal order, stay in many kinds of social system, and inhabit a range of environments according to Clutton Brock in 1989; Fleagle in 2013; Rowe and Myers in 2015; Wright in 1999.

These differences present a vast collection of raw matter that biologists can explore to encourage the general understanding of how morphology, sociality, and behavior evolve under a range of ecological conditions. The precise interest are questions relating to how different species adapt to the same circumstances. For example, research of primate communities, examination of different responses to environmental change and habitat degradation and the way equal species adapts to diverse conditions (For example, documentation of variation in behavior, sociality, and existence records throughout environmental gradients).

Figure 6.6. Bonobo Monkey/

Sources: Image by Pixabay

The extinction of primate species and loss of populations will bog down our potential to make sense of the natural world and elucidate common biological principals that is applicable to many other taxa. The very important role of primates in furthering biological understanding is particularly evident when one considers the ordinary lack of knowledge of the tropics.

The tropics residence the majority of the world's biodiversity (Ceballos and Ehrlich 2006; Kreft and Jetz 2007) and yet for many businesses we lack even the most primary perception of their diversity, biology, or conservation status. For example, in amphibians a lot large proportion of tropical species than temperate species are classified as data deficient by the IUCN (Collen et al. 2008; Stuart et al. 2004), a sample that seems to be actual of other taxa as properly (e.g. mammals: Schipper et al. (2008); birds: Butchart and Bird (2010)).

Against this backdrop of customary lack of understanding about the tropics, primates stand out as an especially well-studied group, in phase due to the fact many species are gregarious, diurnal, and especially effortless to find out about in the wild (Emmons 1999; Harcourt 2000, 2006; Beaudrot et al. 2013).

6.6. THE PRIMATE THREATS

The human activity around the world is putting in danger almost half of all primate species. The basic dangers this group is being facing are habitat destruction, poaching, the unlawful pet trade, and disease.

Status of Feature Species

- Sumatran Orangutan: Critically Endangered
- Bornean Orangutan: Endangered
- Eastern Gorilla: Endangered
- Western Gorilla: Critically Endangered

6.6.1. Loss of Habitat

One of the main dangers, primates are facing are habitat loss and fragmentation. In some regions, human presence into primate habitat causes combat and pushes animals into less suitable areas. In many areas, primate habitat is being cleared for industrial logging and agriculture leaving primates with no vicinity to go. This is particularly the growing issue in south East Asia where land is being logged or burned and cleared to make way for palm oil plantations leaving orangutans with nowhere to go or killed in fires. The Mining in forest areas also poses a big risk to primate habitat. For example, in Indonesia.

6.6.2. Poaching

The bush meat business is a world-wide risk and continues to be a serious risk to primates. Once faraway populations of primates are now greater reachable as their habitat shrinks and humans get reach to the once faraway areas by means of logging roads etc. Though not traditionally a foremost target of the bushmeat trade, this threat is growing for primates such as gorillas as human beings attain entry to their habitat.

Figure 6.7. Central American Squirrel Monkey.

Sources: Image by Flickr

6.6.3. Exotic Pet Trade

There are large numbers of primates that are captured and sold as wonderful pets every year. Regrettably, many of these animals will die during the seize and transport process. Others are condemned to a life in captivity, regularly dwelling in bad conditions. The infant primates are regularly the target of this business and as a consequence, more than one fully grown animal are killed to gain access to them. The illegal business of flora and fauna is a multi-billion-dollar business and is second only to the world black market drug trade.

6.6.4. Infectious Disease

Between the humans and non-human primates there can be possible transmission of diseases. With the growing human settlements, the encroachment on primate habitat, primates are driven into smaller areas, growing their exposure to people as well as other primates, increasing their possibility of contracting infectious diseases. Tourism to primate habitat additionally poses a danger and need to be done cautiously to make sure the protection of primates like gorillas towards infectious disease. The exotic pet business additionally poses a risk to human health as primates can transmit diseases such as Ebola, tuberculosis, and different viral infections.

6.7. WHY PRIMATES MATTER

6.7.1. Social and cultural importance

The primates have a very close evolutionary history with humans, they also contribute importantly to the biological and cultural richness and the natural heritage of the countries in which they are found. There are many primates that contribute significantly in ecosystem dynamics and sustainability and are main characters in local and regional conventional knowledge, folklore, history, and even economies. For example, in the total regions of South and Southeast Asia, monkeys are mentioned strongly in Hindu and Buddhist mythology (for example, the Hanuman langur is the monkey god Hanuman in the Ramayana) and have centuries-old sympatric relationships with humans at temples and holy shrines.

In the present times many of these monkey temples are tourist sites, where the monkeys are protected, and provide an important source of profits for local populations. There are many traditional and old societies that protect or tolerate primates and have combined them into their historic narratives as sacred cultural figures and persons, as pets, and as food. For example, the Amazonian Guajá people combined monkeys into their association and religious systems while additionally depending on primates as a sustainable food sources. The primates are becoming rare or extinct; there is danger of losing complicated ecological, social, and cultural relationships that have developed between people and primates over times.

6.7.2. Ecological Importance

The primates are prey, predator, and mutualist species in food chain and thus influence ecosystem structure, function, and resilience. Their evolution, feeding ecology, and geographic distribution are closely associated to the diversification of angiosperms, an important supply of meals (pollen, nectar, fruits, and seeds) for many animals and human beings. There are most primates that have been recognized or suspected as important pollinators because of their opportunistic non-destructive feeding on vegetation and nectar.

The primates feed on different plant components (for example, fruits, flowers, seeds, gums, and leaves), primates can affect plant propagule dissemination, causing tree mortality, and might negatively affect the reproductive investment of some plant life. Though several primates are especially frugivorous, and their exceedingly large dimension enables them to disperse small and massive seeds over long distances, increasing forest regeneration.

In the non-existence of zoochorous seed dispersal by means of primates, plant populations can experience diminished genetic heterozygosity and growing genetic subpopulation differentiation, growing poor density dependence, and diminished recruitment. For example, Madagascar's lemurs show complicated relationships with giant seed–producing trees, and lemur extinction may also be facilitating a decline in the feasibility of some Malagasy tree species.

The population of primates consisting of large atelids and cebids in heavily hunted forests of Amazonia has severely degraded long-term forest area dynamics and the sustainability of many hardwood tree species with implications for the carbon-storing capacity of forests. In the same way the hunting of gibbons in northern Thailand has had a bad effect on the demography of the lapsi tree (Choerospondias axillaris), which relies upon on gibbons to scatter its seeds into light gaps.

The loss of primate seed dispersers has obvious effects on human populations in the identical ecosystems. The village people in southern Nigeria depend on gathering primate-dispersed fruit and seed species, signifying the largescale importance of primate conservation to local human food security.

6.7.3. Primates as Model Animals

The primates are considered as greatly valued model animals, increasing the knowledge of the evolutionary history of our species and giving perception into human behavior, cognition, parenting, cooperation, adult social bonds, forms of social warfare and resolution, studying and memory, and the evolution of tool use and language.

Though there is presence of important moral issues that want to be regarded when the usage of primates in scientific research, primate models have enhanced our perception of atherosclerosis, respiratory diseases, HIV or AIDS, treatment reaction to psychoactive drugs, psychopathologies, sociality, mental health disorders, communication, immunology, talent functioning, pharmacology, endocrine legislation of reproduction, genetics and genomics, and disease danger and parasite dynamics, amongst many other subjects.

The wild primate populations might hold important clues to the origins and evolution of important pathogens and approaches of natural disease transmission through serving as guards for early disorder detection, identification, and surveillance, accordingly, benefiting humans. The growing infectious illnesses also pose serious threats to both endangered and no endangered primate species, research of these illnesses in one primate population can also benefit conservation efforts for others.

6.7.4. Addressing Conservation Needs

The global scale disadvantage of primate fauna is caused by deforestation, hunting, unlawful trade, and timber extraction. Drivers of primate loss are dynamic and engage with every other at local, regional, and world level, leading to a path of bio-simplification that is most keenly felt as marked decrease in population sizes and, all too soon, extinctions.

The world scale of primate population declines and the predicted growth in the intensity of predominant anthropogenic threats endorse that conserving wild primates is an instant but daunting challenge. The lack of huge systemic changes in human behaviour, populations will proceed to decline over the next few decades, with species presently listed through the IUCN as endangered becoming extinct and species now categorized as near threatened or least concern at the verge of increased extinction risk.

Figure 6.8. Spider Monkeys.

Sources: Image by Flickr

The most of primates are iconic (for example, gorillas, chimpanzees, orangutans, spider monkeys, and lemurs), though given the level of their decline, it is clear that neither their appeal nor their flagship repute is sufficient to protect them from the threat of human-induced extirpation at some level in their native ranges. The extermination not often effects from poor scientific understanding of the steps needed to guard the species. In its place, it is entrenched in political indecision, socioeconomic instability, organized criminality, corruption, and policies that give preference to short-term profits over long-term sustainability. The good primate conservation will need a principal revolution in commitment and policy. The removal of pressures upon primate habitats needs decreasing the per capita demand of industrialized countries for tropical hardwoods, beef, palm oil, soy, rubber, minerals, and fossil fuels, among different goods, whilst concurrently advertising sustainable resource-use practices.

6.7.5. Improving the human condition

The human capital in primate regions localities is of greatest significance to primate conservation. The increased rates of human population increase (5.1 billion human beings in 2010 to 7.3 billion in 2050 in primate range countries), high scale of poverty and inequality, the loss of natural capital because of large scale and fast land-cover alterations driven by means of international market demands, poor governance, and the need for food security, health, and education are basic things to think about.

The answers to the problem of primate conservation must contain lowering human birth rates and population increase, betterment of health, decreasing poverty and gender biases in education, creating sustainable land-use initiatives, and maintaining conventional livelihoods. At the local level, human beings and governments need to turn out to be partners in this effort and thought that they can gain from defending primate populations and their forests while at the same time fulfilling their primary needs. It will need education, rethinking, and investment from government, nongovernmental groups, and the private sector businesses.

6.7.6. Expansion of protected area

Though the percentage of the land surface dedicated to protected areas has gradually increased in primate areas countries, highlighting covered areas as the solely plausible conservation tool may make contributions to local region poverty by denying poor people control over and get access to the natural resources that support their livelihoods.

The protected land regions require being sufficiently massive and providing appropriate primate habitats, and species of apprehensions need to be existing in sufficient numbers within these areas. Regrettably, the counties in the Neotropics, Africa, and Asia are presently downgrading, downsizing, and denotifying covered areas because of developing industrial-scale natural resource extraction, and the extensive number of protected areas is facing sizeable deforestation.

In spite of these trends, protected areas can and do give long-term sanctuary for wildlife. For example, the research in the Udzungwa Mountains of Tanzania confirmed that colobine primate populations were steady in the protected areas however declined severely in the unprotected forests. The eight research, the use of camera traps to yearly screen terrestrial mammals and birds in 15 protected areas in the Neotropics, Africa, and Southeast Asia presents strong proof of balance and also growth in populations, include of those of 23 primate species 19 cercopithecines in Africa and Asia, three African apes, and one lemur.

6.7.7. Land-sharing and land-sparing

The forests are the few financial resources that are available to the rural area in the tropics, securing their possession and balanced commercial use can support poor families fight with and come out of poverty. In a land-sharing strategy (mixing protection and manufacturing in an agroecological matrix), community-managed forests are one alternative to combine forest management into national poverty decrease programs in rural areas, favoring primate species with small region's needs.

These forests also have lower and less variable annual deforestation speed as compared to that of protected ones. In both land-sparing favoring species with broad regions needs and land-sharing approaches, promoting biodiversity and the utilization of lattice-work corridors to join landscapes along latitudinal and elevational gradients may additionally promote a variety of habitats for the long-term persistence of primate species that differ in their ecological necessities and may additionally mitigate some of the deleterious effects of climate change.

6.7.8. Use of New and Traditional Technology to Monitor Primate Population Vulnerability

The latest way of observing primate habitats, population status, and anthropogenic threats is presently emerging. It consists of taking benefit of international telecommunication systems and Wi-Fi Internet, satellite- and airborne-based imagery, drone technology, ever greater effective handheld gadgets (for example, smart phones and tablets), and camera traps.

Integrated with geographic information system and ground surveys, some of this technologic has been utilized in assessing sustainable land-use spatial planning and human-primate conflicts for example, Javan gibbons (Hylobates moloch) and in supplying case-by-case assessments of species vulnerability to climate change.

These same technologies can also be used by local citizen scientists for species and forest observation, therefore bettering the effectiveness of mitigation measures. The latest technological developments in molecular biology specially high-throughput sequencing of DNA extracted from non-invasive samples (for example, faeces, urine, and hair)—can speed up assessments of population measurement and structure, genetic diversity and proof of out breeding, diet (plant, vertebrate, and invertebrate DNA consumption), and parasite and gut microbial range for wild primates.

6.7.9. Mitigating unlawful trade

It has been observed that to lessen natural world poaching, interventions need to go past law by encouraging capability building in local communities to protect wildlife, re-examining sustainable off take mechanisms, such as regulated trade, ranching, and natural world farming, and to use social media and the Internet to decrease demand and, finally, to limit business.

Though the utilization of social media to increase local, regional, and international focus of the plight of the world's primate fauna and of the ecological, social, cultural, and financial significance of primates is, no doubt necessary, it is just as necessary to enhance local, action-oriented conservation education programs, mainly those concentrated on young people and community decision takers.

6.7.10. Reintroductions and Long-Term Forest Area Safety as Conservation Tools

The areas where primate species are no longer found locally, reintroductions are a choice however can be expensive, and long-term safety of forests is arguably a greater cost-effective way of conserving primates than reintroduction. On the other hand, reintroductions increase public and political focus and offer placement solutions for rescued animals in line with welfare concerns. The use of wild-born, rescued, and rehabilitated primates as an alternative of captive-bred animals in reintroduction programmes decreases expenses and can give more success.

6.7.11. Decreasing the city Footprint on Primate Habitats

Around the world, urban area policies need to be centered at decreasing people's ecological footprints in primate range areas. For example, promoting the recycling of cell phones, laptops, and other digital devices could decrease the demand for coltan mining from the Congo Basin in Central Africa, which threatens primates in the region that includes gorillas and chimpanzees.

The reduction in the world's per capita demand for tropical hardwoods, foods and non-food products, minerals, and fossil fuels, along with different goods, from primate range areas would facilitate in lessen pressures on primate habitats.

REFERENCES

1. https://news.janegoodall.org/2017/02/27/hope-primate-conservation/ [Accessed 27 Aug. 2019].
2. https://www.nature.com/scitable/knowledge/library/primate-conservation-is-the-cup-half-empty-53872551/ [Accessed 27 Aug. 2019].
3. Estrada, A., Garber, P., Rylands, A., Roos, C., Fernandez-Duque, E., Di Fiore, A., Nekaris, K., Nijman, V., Heymann, E., Lambert, J., Rovero, F., Barelli, C., Setchell, J., Gillespie, T., Mittermeier, R., Arregoitia, L., de Guinea, M., Gouveia, S., Dobrovolski, R., Shanee, S., Shanee, N., Boyle, S., Fuentes, A., MacKinnon, K., Amato, K., Meyer, A., Wich, S., Sussman, R., Pan, R., Kone, I. and Li, B. (2017). Impending extinction crisis of the world's primates: Why primates matter. *Science Advances*, 3(1), p.e1600946.
4. https://sites.lsa.umich.edu/wp-content/uploads/sites/162/2016/07/Marshall-Wich-2016-Why-conserve-primates-small.pdf [Accessed 27 Aug. 2019].
5. https://www.neprimateconservancy.org/primate-conservation.html [Accessed 27 Aug. 2019].
6. https://seethewild.org/primate-threats/ [Accessed 27 Aug. 2019].
7. https://www.researchgate.net/publication/233981144_Primate_Conservation_Efforts [Accessed 27 Aug. 2019].

Chapter 7

Challenges in Primate Conservation and Primate Biogeography

"One of the great commandments of science is, 'Mistrust arguments from authority'. (Scientists, being primates, and thus given to dominance hierarchies, of course do not always follow this commandment.)"

— Carl Sagan

CONTENTS

There are several challenges which are counted under the primate conservation. This chapter provides information about these challenges in the primate conservation which are properly described in it. In the first half, this chapter provides insights about the rehabilitation and reintroduction of the endangered species and the challenges which are in the captive breeding. This topic of challenges in captive breeding includes the reintroduction of selected primate species in the study area, views of respondents about the captive breeding, general problems which are affecting the operations of captive breeding and the Gibbon rehabilitation. There are some other factors which creates hinderance in the conservation action have also been mentioned in this chapter. In order to study the biogeography of the primate conservation, the description about the molecular clock studies have also been discussed in the end of the chapter.

7.1. INTRODUCTION

There are two major considerations that drives conservation decisions which are whether a path to extinction that exists under the current circumstances and in case, if it is feasible then it alters that path. About 50 percent of the 600 extant species and subspecies of primates are at risk of extinction. In 20069, the four of the most endangered primates were from Madagascar, seven of them were from Africa, 10 were from Asia and the four occurred or had lived in the neotropics. The population of the Gorilla and chimpanzee have been decimated by around 80 % in the Congo Basin.

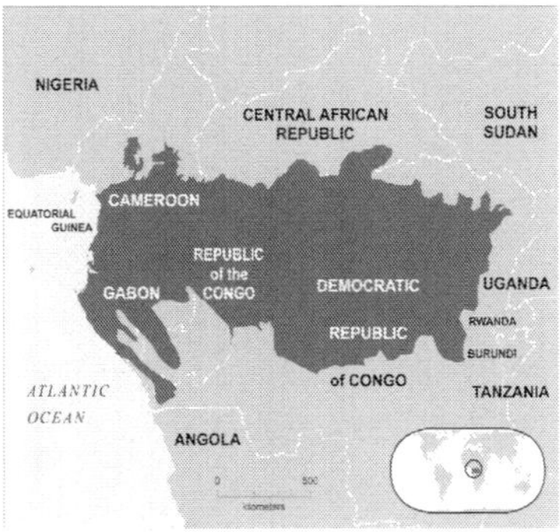

Congo Basin Wilderness Area

Figure 7.1. Congo Basin Wilderness Area.

Source: Image by Wikimedia

Around one third of the overall Madagascar species were extinct by 1997. Therefore, it is clear that a large number of primates are on the direction towards extinction. Despite this, the most important question that arises is whether that path can also be altered or not. Once it was thought that sufficient genetic diversity can be maintained by retaining a population with only 500 individuals. Then the estimate was subsequently revised upward to 5000 individuals.

There are certain primary factors that directs the path to extinction are anthropogenic environment alterations such as logging, direct targeting like poaching, climate change and interaction between humans and disease. Deforestation mostly finalizes the extinction of populations which was initially decimated by hunting as well as logging.

The major challenge that needs to be addressed is the habitat loss. The impact of such a challenge is variable which can lead to either to the change in behavior or extinction. The former often places the primates in conflict with their human relatives. Considering the deforestation, then it as greater than a 90 % depletion of the closer – canopy forest tree cover. Africa experienced a loss of 10.5 % in the period of 15 year in between 1980 and 1995; Latin America and the Caribbean, 9.7 % and in the Asia and Oceania, it is 64 %.

There are certain economic advantage of timber harvesting and agriculture work and this is against the conservation. The deforestation has led to the 1,23,000 tons of the primate biomass loss per year. In case, if it is accessed in terms of edible meat, then these populations have an estimated value of $146 million dollars.

While considering the extinction, the forest loss can be better tolerated by smaller primates but those which are with dietary dependence on frugivory are less tolerant. After the burning of the forests, a few numbers of primates can persist for example, the marmosets and titi monkeys (Callicebus hoffmannsi). The primate extinction can also lead to the deforestation. The dependency of the regeneration of trees is on the seed dispersal by primates. Those seeds which just fall to the ground have a low survival probability.

Figure 7.2. Red Titi Monkey.

Source: Image by Flickr

Climate change has a major impact. The galada baboon that is Theropithecus galada, represent the species vulnerability. The lower limit of its altitudinal range rises 500 meters with each 2-degree centigrade increase in the average temperature. This will halve as well as fragment the range which is available for its habitat. The response can be neither immediate nor obvious. According to the Struhsaker, there is a decline in the vervet monkeys Chlorocebus aethops which was delayed for around 10 years after the 90 % of the loss of their major food resource.

According to the Struhsaker, there is a decline in the vervet monkeys Chlorocebus aethops which was delayed for around 10 years after the 90 % of the loss of their major food resource.

Along with the primate species comfort zones, the effect of the warming on the foods supply of the primates is very more important. There can be alterations that can be done in the plant species range, character of leaves, reproductive and transmission habits such as flowering or fruiting. As per the suggestions of the McClean, four of the current five thousand African plant species will need to be migrated to some high elevations and further suffer variable degrees of habitat loss having more than a quarter suffering of the total loss of the habitat by 2085.

There are both direct and indirect consequences of the proximity with humans. When the monkeys develop a taste of the human foods and lose fear of the humans then that leads to the increased interaction by the crop raiding. There are risks such as the bites, non – provisioned food theft and simian viruses and furthermore diseases which is often related to the ecotourism.

The humans share a susceptibility to the disease with other primates which includes parasites such as Bastocustis hominis and Trichuris trichura. Fruits which are newly popular with humans include the traditionally consumed by some other primates. Further, this has led to the reduction in the availability for the latter. Primates have also suffered predation from half the dozen of the domestic dogs which are typically owned by the rural families in the Peru. One more challenge in relation to the survival derives from the competitive pressure on the terrestrial vegetation which is created by the livestock.

Figure 7.3. Hairless Peruvian dog: Peru has a race of dogs that have no hair at all, apart from the top of the head.

Source: Image by Flickr

7.2. REHABILITATION AND REINTRODUCTION

Considering the world's primates, the majority of their population live in the third world countries that is those countries which are experiencing the speediest increase in the human population and so, it exerts a huge pressure on their natural resources.

Despite this exposed situation, the unexpected opportunities can arise which can possibly because of the greater likelihood of the public identifying with, and furthermore providing the financial support to, which is familiar and flagship species in order to establish the semi – natural habitats which can be used for the purpose of reintroduction. In such context, we can take example of the USA. In the case of marked bob – white quails, peregrines falcons and various insects, fishes, amphibians and reptiles.

Keeping this in mind, the best action can be to try the reintroduction to as many endangered species as possible in an economic manner, along with this, focusing on the more popular birds and mammals. Only half of the reintroduction shows that it has released threatened as well as endangered species or sub – species. This shows that the potential for the R&R has yet to be realized other than some encouraging success stories.

The immediate conservation of primates becomes very much urgent as suitable habitat and the number of primates are reducing day by day. The plight of the primates is still an open and popular topic for the debate, despite that the endeavors of the previous rehabilitation efforts have either improved or failed so as to have any impact.

Even though, the importance of the reintroductions and rehabilitation programmes cannot be overlooked with regards to the community and international education, the awareness has been increased and the refuges for the abandoned animals. In case, if the animals were not brought to rehabilitation centers, they will face some uncertain future in some unsuitable conditions.

7.3. CHALLENGES IN CAPTIVE BREEDING

7.3.1. Challenges Faced in the Captive Breeding and Reintroduction of Selected Primate Species in the Study Area

The respondents observed that there were alterations in the behavioral patterns of the released primates because they were bred ex – situ. They become unable to interact with the wild ones and was unable to cope with the forest situation because of the fact that they have lost their foraging skill and the ability to escape from the predators or hunters.

This shows that they have lost the sense of wildness. The released primates are returned to the captive breeding center and this cannot be unconnected with the fact that the primates are sure of the free food in the breeding site of which they are used to. This can be endorsed to the fact that when the species are in the breeding sites, they were not properly hardened up so as to be used to the wild.

In addition to this, the breeding sites were not spacious enough so as to serve as the home range of the species in the wild and the primates were not raised so as to run away from the humans. This is in the agreement with the work of Dellatore. In the challenges indicated by the respondents include the stress and time that is taken in order to raise a self – sustaining population.

There are a large number of primates which spend their lives in either large groups or communities because this helps them in order to provide the protection against the predation and furthermore, protect their scarce food resources. Each group of the primates can consist around 25 in number and thus, composed of few males and several females with their young ones which include every member acting as a team. This population ratio can be attributed to the source of breeding stock as well as making the selection of a number of various groups for the re – introduction much difficult.

Figure 7.4. Captive Breeding – Red Wolf Pups.

Source: Image by Wikimedia

There is requirement of time in order to raise this kind of group which will be fit for the re – introduction, because it requires close monitoring of the ones which can understand themselves in the breeding stock, after that separating them from the breeding stock so as to increase their understanding. In the situation of preparing them to get released, constant screening, feeding and behavioral monitoring will be going on along with the increasing release of stock.

In case, if the ratio of the males to the females is very close, then there will be a need to remove some males leaving the infants and other females in order to avoid the fighting for territoriality and further prevent inbreeding. For doing all of this, the time and other resources are spent. According to respondents, they have released just once in 2007 and the stock was just three in number, since the inception of the project in Cercopan.

The group was not a self – sustaining one and so they all returned to the breeding center. In such cases of Pandrillus, the respondents showed that the re – introduction has not been embarked upon since the inception of the program that too in the year 1991. Therefore, this shows that it takes time in order to raise and select a self – sustaining group that is to be released.

This agrees with the observation of Clark that is the source population of primates can be in short supply, specifically if the target is a certain age or sex groups or if the source contains the endangered stock. If these organizations are really interested in entering into the successful reintroduction exercise, then they will have made a number of attempts.

The Cercopan did not make any other attempt in order to reintroduce the animals after failing and the Pnadrillus did not attempt any re – introduction exercise. The fat behind this can be an indication that the intention of the organizations can be so as to manage the captive bred animals in perpetuity.

This demonstrates the findings of Hancocks as per citation done by the Laidlaw that the most conservation centers around the world create so much awareness about the captive breeding and reintroduction programs, despite this, only a few participate in actual in them in a substantive way.

Even a number of the high – profile Species Survival Plan (SSP) breeding programs in various parts of the world, have no mechanism for reintroduction and all of them concentrate on maintenance of captive populations in perpetuity. According to the observation done by the respondents, the other challenges is the issue of disease transmission.

During the research, as per the observation by the respondents, they are in contact with primates daily and they do so without any form of protective gadget like the hand gloves and face masks. This kind of close contact carries a considerable risk of exchange of zoonotic diseases similar to those carried by humans that can spread to the primates and vice versa.

The spread routes can be through aerosols (cough and sneezes), water contamination, bites etc. The respondents also complained about the bacterial diseases as the ones having highest occurrence in the vicinity and therefore, they can be transmitted through a number of different means.

Respondents complained about the long duration and stressful nature of the monitoring periods and the inadequacy of the monitoring materials. This can be attributed to the fact that the primate species in the Cercopan were released during the dry season in order to ease the accessibility to site since the staff has to keep on monitoring the releasing of the stock until the season of rains starts and this is when the fruits become abundant in the wild.

As per reported by the respondents, this exercise is done manually with the crude implements like the working down to the release site without any protective in order to supply food to them on daily basis. In the process of monitoring, some of the staff stay overnight with the released stocks in order to observe and protect them. This puts the lives of workers at risk.

In the case of Pandrillus, since the inception of the project the re – introduction has not been carried out. This can be attributed to inadequate protective measures, lack of monitoring materials because of the high cost of purchasing them, the lack of the skilled personnel and the health requirements of the primates.

There was an interview done with the vet doctor. There, in that interview, he complained about the impotency of the locally purchased drugs not similar to the ones which are imported from other countries that trend to be more potent.

Some of the respondents complained about the attitude of the community towards the project. Despite the high awareness level of the villagers about the conservation project, there are cases of poaching which are still noticed as villagers who still encroach into gazette site in order to hunt and carry out logging operations.

Even though the positive overall attitudes toward primates, the attitudes of villagers towards primate re – introduction have been discouraging workers. These negative attitudes are associated in a partial manner with the perception that the re – introduction will lead to restrictions in the land use. Rural property – owning and resource – dependent groups tend to be more utilitarian and dominionistic in their value in the direction of wildlife.

In similar to this, most of the workers are of the view that successful re – introduction of the primate can lead to the end of the conservation project and losing of their jobs. Therefore, they perceive the captive management of these animals in perpetuity will provide security to their job and attract sponsorships. This is an attitudinal issue which is in line with the work of Clark and Kellert.

7.3.2. General Knowledge of respondents about Captive Breeding and Re-introduction of Primates

There is a large number of the respondents from the both the conservation institutions which are not conversant with the processes which are not involved in the captive breeding and the reintroduction of the species. This can further be described in order to the love the level of educational qualification. The people which are more educated, tends to learn faster in comparison to the less educated people. This represents both the institutions which are using the same strategy.

This also represents that the institutions have not been given relevant trainings on the techniques which are involved in the breeding and re – introduction of primates into the wild. The implication of this is that the suitability of the program can be guaranteed hardly as the techniques for the program are known by the smaller number of persons.

It can also be said that a small number of persons who are a part of the management team, do want to remain relevant in a sustainable manner and maybe they are not willing to train others so as to acquire the skills which are required. And therefore, this can prevent the large number of the workers from understanding the processes. These are the processes which are involved in the breeding and re – introduction exercise.

7.3.3. General Problems Affecting Efficient and Effective Operations of Captive Breeding by Non-Governmental Organization (NGO)

As the problems of the conservation institutions, the unavailability of fund, poor staff health and welfare packages, poor salary scale and the lack of infrastructure has the highest cumulative percentage. The reason being the inadequate fund, the organizations are becoming unable to purchase the standard equipment like the monitoring camera which works with the satellite and patrol vehicles for the job. This creates hindrance in the organizations from employing more staff because the already existing ones are not well remunerated.

The organization contributes very little to the hospital bills of the respondents when they are sick. A respondent complained about the fact of not earning up to ten thousand naira even with his years of experience as well as family status. In addition to this, there is no insurance which can cover them, in case they are injured along with carrying out their duties, whereas they are exposed to a number of different kinds of risks.

In the case of infrastructure, the observation of the roads was done and seen to be in a bad condition as sometimes, the researchers were determined to drop and walk across the water bodies prior to accessing the site. The lack of transport related facilities makes the workers travel long distances on the daily basis prior to getting to the site to commence work. This led to the limited home range and no release of captive – bred individuals to the population of animals which has increased the carrying capacity of the breeding centers. This resulted in the problem of inbreeding and consequential increase in the chances of disease infestation.

This is in the agreement with the works of Clark, Reading and Kellert. Along with this, 79.17 % of the respondents in the Pandrillus registered complaints about the loss of habitat because of the deforestation which is the result of some human activities like the bush burning, logging and fuel wood collection.

Figure 7.5. Logging Then and Now.

Source: Image by Flickr

Despite the fact that the villagers are conscious about the project, they can still carry out deforestation activities in the form of bush burning in order to shift cultivation and burning of bush meat, clearing for fuel wood etc. This endangers the efforts of these organizations as the habitats of these primates which are destroyed by these activities.

A similar report was provided by the White and Tutin says that the logging practices were responsible for the decrease in the chimpanzee densities in Gabon. Competition is created among the animals due to the provision of food to animals especially as the food which are more concentrated at locations and always provided unlike in the natural forest where they are dispersed and furthermore are available mostly on the basis of season.

The challenge is the competition during feeding and so, occurs because of the fact that primates observe the principle of peck order that is social stratification. Provision of food can result in the increase in the proximity and aggressive competition.

Primates are considered as the higher - ranking individuals which chase away those which are in the lower classes when being fed which can lead to a serious fight that can result in the injuries. While some tend to escape from the enclosures without considering the risk of being electrocuted. The results which are similar to his are obtained by the Berman in 2007. The workers in these organizations are very few in numbers and majority of them have secondary level of education irrespective of the skills which are required for the type of job that needs to be done.

This can also affect the efficiency and furthermore obstructs the free flow of the activities in the project. This causes inadequacy of staff which contributes to some form of the insecurity of the protected site similar to poachers like the loggers and hunters gain access into the reserve. The fact that most of the workers are secondary school leavers which further indicated that the firms lock professionals and the few who know the job, refuse to teach others about the same.

Further, this indicates that there is no transparency and free flow of the ideas in the staff of the organization. Besides this, the equipment used are obsolete and few. There are only few workers which had safety booths and all the workers do not had safety coats. This reveals that the vulnerability level, to which the workers are exposed to, are high.

7.3.4. Gibbon Rehabilitation (applicable to many other primate species)

While the social structure and behaviors of gibbons impede them from being used as a direct model for other primate R&R programs, the following lessons should be well known as is best applicable for other primates as well.

It is usually seen that there are many problems that gibbon rehabilitation faces: some of the gibbons have spent their entire life in tiny cages, chained and drugged, while others will have been comparatively treated very well and may face complication in adapting to life without their human family (Cheyne 2006). Tame gibbons form fewer stable pair bonds, and their usually shows unpredictable behavior towards conspecifics (Eudey 1992).

It is important to gain an insight into different types of stereotyped behavior that captive gibbons generally possess and is very important in case efforts are made to improve husbandry techniques that minimize or eliminate the causes and symptoms of poor welfare and the resulting central nervous system dysfunctions associated with stereotypy (Cheyne 2006).

Research should be dome inn such a way that it possesses the ability to differentiate between gibbons that are orphaned during their younger age, and those that have been kept for a certain period as pets, as their care and rehabilitation needs will be completely different. The lack of efforts to take into consideration the different life histories of the captive-raised gibbons has been an oversight in the past (authors' pers. obs.).

Non-human primates in the wild have been raised with proper parental care with long periods of infant and juvenile dependency (Yeager 1997). During this life stages, the youngsters learn many behavioral, social and sexual responses, e.g. how to communicate with others, how to avoid predators and how to manipulate and handle food.

Grouping juvenile gibbons is very helpful for socialization as well as for the identification of compatible individuals; however, only bonded pairs — and not groups of juveniles — should be released.

One of the other option available is to put an older, greater than four year older, individual with an adult companion. This substitute parent idea has applicable between 2 white-handed gibbons Hylobates lar males (author's pers. obs.), but it is still not clear whether it is not applicable in case of all animals or whether these 2 were unique with respect to their ability to co-exist despite not being related.

Still now, there is no information available about ability of these males to form a pair bond with a female. The possibility of an adult gibbon teaching a youngster is far more preferable as compared to the human teaching the gibbon the relevant level of survival skills.

However, there is one downside of this: in case the juvenile and adult are not compatible, then it may result in juvenile developing a fear of conspecifics, causing severe problems during the time of formation of pair bond. The increase in the numbers of gibbons that have been brought to rehabilitation centers has forced to shift emphasis from rehabilitation to simply trying to find humane housing for the newcomers.

As a result of this, there has been very minimal focus on the rehabilitation aspect, and until this is efficiently solved, the centers will consistently fill up with animals that face an indeterminate period in a cage. Some of these gibbons have spent majority of their life being free to roam (albeit in a human environment) and may not adapt well to being caged.

In case if it is early predicted that an animal is ready for potential reintroduction, there is necessity or compulsion to give this animal a chance at freedom. Other major problem is that while a gibbon pair may have copulated and duetted in captivity, this is no sign that the association will continue once they are released. The pairs may split and re-form, an individual may fall ill or may disappear or not adapt well in the prevailing condition and have to be brought back to the center.

The primary cause of failure should be devoted to a lack of information about the species, the release area and the individual's action to the release area. The animals should be monitored and studied well over the long term to evaluate and assess their reactions in the new environment and to record in case of any deaths or disappearances. In case the released animals do not live long enough to breed, then their conservation value is negligible or infrequent (Cayford & Percival 1992).

Although there are many documented programs for rehabilitation of primates (Wynhoff et al. 2000, Tutin et al. 2001, Farmer & Jamart 2002, Cheyne 2005, Goossens et al. 2005, Farmer et al. 2006, Cheyne & Chanee 2008, Cheyne et al. 2008a), each species usually possesses a unique challenge for successful rehabilitation or reintroduction. Problems that the gibbons face in R&R projects can be summarized and discussed below:

- Developmentally stunted social skills
- Possible stereotypic behavior

- Malnourished and/or unable to brachiate properly
- Unable to learn or sing the duet, the species-specific morning singing about where the male and female are involved in singing sex-specific songs.

7.3.5. Other Factors Hindering Conservation Action

Respondents identified different number of factors that get in the way of effective action to address these threats. These comprises of lack of political will to deal with the issues, often accompanied by corruption and/or the existence of political systems that allow powerful people to manipulate the law for their own benefits.

It generally include the practice of overriding the laws; limited funding for conservation; deficiency in scientific knowledge, including the present or current information on population distribution patterns and abundance; and lack of adequate policies that exaggerate economic development in conservation strategies and which result in top-heavy administrative structures in conservation organizations.

This set of possible threats, hindrances and drivers to action is consistent and is gradually increasing. There is some other consideration which are important in this subject matter. The first is that present activities of primate conservation are far from evenly spread all over the world and across different species.

For understandable reasons, certain taxa, e.g., great apes, and certain regions, such as Madagascar, receive greater attention compared to others, with the result that some highly threatened and endangered once primates, such as many forms of red colobus monkey, continue to be ignored and neglected. Second, conservation donors, organizations, and others often deploy larger amount of resources to meetings and workshops about conservation, rather than initiating direct action to protect threatened primates as well as other wildlife.

7.3.6. Recommended Actions

What are the possible ways to address these challenges effectively? The views or opinions of many are summarized in this section. It is worthy to note that a large majority of the elders felt the need for more awareness-raising about the importance of conservation.

To summarize this in one line, it can be concluded that there is a need to change the hearts and minds of both the rich and the poor, so as to aware them about the consequences of their activities and actions, change their perception and attitudes to nature, and obtain greater political support for conservation.

Among specific actions mentioned above, the use of broadcast media to a greater extent in primate habitat countries plays very important role and the use of charismatic personalities to deliver conservation messages.

The second most important thing is to recognize increased awareness-raising as a conservation need, most people have argued for the necessity to make protected areas more effective, through promoting the enforcement and vigilance of genuine law (in which, it was pointed out, dedicated management can play an even more prominent role than greater funding), while seeking support both from central government and from surrounding communities, including measures to improve their well-being.

Among other actions that were suggested by respondents, there were various other individuals who argued the need for enhancing the population survey techniques and knowledge of the basic natural history of poorly known species.

With the opportunities to make new protected areas becoming restricted in the face of an increasing human population and growing human demands, two respondents outlined the value of introducing more biodiversity conservation into agricultural landscapes in the developing nations, as happens widely in the region of Europe (this is not a strategy that functions well, however, in the scenarios where large mammals that raid crops are existing).

In addition to the obvious requirement to raise more money for the purpose of conservation, among other points raised were the significance or need of training and comprising more primate habitat-country nationals as well as the value of field research stations. Among all, one of the respondents pointed out that efficient conservation needs the establishment of honest, trusting relationships among the stakeholders.

There were many experts or scientist who have stated about the importance of making protected areas more efficient and of the need to alter the attitudes in favor of nature conservation, and there were many other experts who laid the emphasis on the same points.

What, then, has gone wrong? Why there exist fifty per cent of primate species that are facing a high to extremely high risk of extinction? Have primate conservationists been suggesting correct or inappropriate actions, or have appropriate actions been suggested but not efficiently executed? If the latter is the issue, what has got in the way of efficient action? Have too few resources of money and people been applied to producing the wanted outcomes? On the other hand, as some elders recommended, are the pressures now being applied by the individuals on the natural world just so great that declines and extinctions of animal populations and species are inevitable?

Though many of the individuals are unsurprisingly quite gloomy regarding the destructive trends that are apparent in the world. There are many individuals who still seem to feel that all is not lost, and that if the correct steps are taken, a majority of extant primate species and subspecies can be conserved. If the individuals are correct, then it is worth trying to identify what should be done in order to attain greater success.

7.4. CHALLENGES IMPOSED BY TIME

Time constraints can restrict the potential of an animal to survive in a specific habitat and the maximum size of its group. Therefore, there are many studies which have examined the ecological correlates of time assigned to travelling, vigilance, and foraging.

However, animals spend more of their time being inactive than active and identifying the determinants of this resting time may offer new visions into the habitat-specific time-budgeting difficulties that the animals face.

The environmental constraints are analyzed that determine the minimum amount of daytime an average primate has to spend for the purpose of taking rest, making use of data from a wide range of ecologically diverse species.

Though, total resting time includes two components:

- Enforced resting time: The time that is imposed on the animal by ecological constraints and
- Free resting time: The time that is available for allocation to ecologically functional activities.

It has been shown that ecologically significant enforced resting time is determined by diet and annual temperature as well as by the variation in the temperature. The tests of the biological significance of this relationship depicts that enforced resting time differs among locations that are suitable or unsuitable for specific genera.

In addition to it, it has been shown that an annual temperature increase of 2–4 C would greatly increase enforced resting time. This give rise to serious time-budgeting problems for many species. The impact of variations to enforced resting time on the biogeographically distribution of species is especially strong for folivorous primates. This study represents that resting time is considered as a significant component of animal behavior that can help in understanding the extinction risk as well as geographical distribution of taxa.

Figure 7.6. Alouatta palliata - Folivorous Species.

Source: Image by Wikimedia

Time is considered as an important resource for all animals, and decisions regarding the time allocation have always been at the center of optimal foraging theory. At the time of feeding, travelling and being vigilant have attracted considerable attention as activities that have instant effect on fitness, resting time has often been seen as being simply time that is not needed for more significant behavior.

Though, resting reflects not only time waiting to be allocated to something more valuable, but it also reflects the time that is required for predator avoidance, recuperation, digestion as well as thermoregulation. In addition to it, it can be considered as a significant variable that have impact on the primate group size.

Therefore, it is considered as significant to distinguish between at least two separate forms of resting that together add up as total resting time: enforced resting (for the purpose of digestion and thermoregulation) and uncommitted resting time.

Only the second is present for the purpose of conversion into additional feeding, travel or social time when these are needed. Enforced resting time may impose important restrictions on the distribution of species by making it problematic for animals to assign sufficient amount of time to other biologically essential activities.

This may be particularly difficult for diurnal species that are not able to compensate by foraging at night for time lost because of high heat loads during the daytime. In principle, enforced resting time is most possibly to be influenced by climate as well as by dietary considerations.

Species that are living in extreme climates may be obliged to search for shelter or simply decrease thermally costly behavior, either to decrease the rate of heat loss at high latitudes or to minimalize heat load at lower latitudes.

In the tropics, for instance, thermal stress that is resulted from high ambient heat loads and high radiant heat gain over the middle hours of the day results in a significant increase in core body temperature. The vertebrate brain is particularly sensitive even to fairly minor variations in tissue temperature.

Though many of the species have anatomical mechanisms to decrease the heat load (for instance, venal rates in order to dissipate blood heat, enlarged muzzles for the purpose of increasing evaporative cooling), searching for shelter and resting is often the most efficient behavioral solution.

The other factor that have the impact on enforced resting time is likely to be the demands of digestion (which, in turn, will partly reflect diet as well as quality of the food. This is most obvious in the case of ruminants, which are forced to spend considerable amounts of time inactive to permit rumination.

7.4.1. Current Issues

The period of intense research and synthesis has been started for primate biogeography. There is so much interest that is on the recording of the spatial patterns of species in the world today and alterations changes that can be reproduced from records of the past with an end goal to distinguish past patterns and anticipate designs for the future (e.g., Wright and Jernvall, 1999). The mankind capacity to recreate the procedures that drive primate biogeography hinge heavily on our learning of a few essential, yet ineffectively parts of primate biology:

- The individual primate taxa dispersal capacities and the components that impact these capacities. All primate taxa (aside from potentially people) are encompassed by zones where the species can't keep up a population due to various physical conditions or a shortage of required assets (Fortin et al., 1996; Legendre et al., 2002; Fagan et al., 2003; Bauer and Peterson, 2005; Stevens and O'Connor, this volume).

These dispersal restrictions might be associated with different types of topographic and environmental situations. The species can be restricted in their circulation by topographic obstructions is its example. Therefore, ranges of mountain work as good barrier as their altitude can provide conditions like that is very cold for most primates that have adjusted to the hotter conditions found at lower heights.

The physiology of primate species is the ultimate obstruction, which will in general be adjusted to a constrained scope of natural conditions. The multidimensional niche concept (MNC) is used by researchers at the time of discussion about dispersal examples and constraints of a species.

According to Hutchinson in 1957 the MNC is a theory that describes how various environmental causes limit abundance and distribution. Since every species has a scope of tolerances and inclinations along each niche axis (habitat, diet, precipitation, and so forth.), a species can only happen in those regions where niche axes are within scopes of tolerance.

According to Brown in 1995 Population increasing rates are on peak where the highest number of niche axes is nearest to most ideal conditions. Hence, there is extraordinary enthusiasm for finding the way the physical restrictions of a primate species and environmental slopes contact to create historical and ecological examples of dispersal.

- Remaking the history and impact of unsettling influences upon primate taxa. The Reconstructive research place our comprehension of distribution patterns of primate species, and the environments they run into, in a context of history. In biogeography, an aggravation is any natural or human-related procedure that upsets the structure or composition of a habitat type. The impacts of the disturbance can be either transitory or permanent.

It is useful to divide disturbances into two classes

(1) Those that impact the structure of an ecosystem, and

(2) Those that impact primate society structure inside the habitat, region, and or environment.

For instance, primatologists will in general pay attention how anthropogenic disruptions (e.g., fire, logging, and human land-clearing exercises) impact primate societal structure. The extinction can be impacted by and extirpation patterns of some species.

The impacts of ecological disturbances like flooding regimes, tree falls on primate community structure have hardly been studied though the anthropogenic disturbances are regularly referred in debates of primate biogeography and conservation.

7.4.2. Discussion

In order to fully comprehend the way, the time-planning issues influence primate survival, the experts have recognized the environmental causes that determine how much time animals are compelled to rest. It has been proven that the level of leaves in the diet, temperature and group size affects the overall resting time. This impact is to a great extent free of phylogenetic relatedness.

The enforced resting time, in any case, sets limitations on biogeographical dispersions because, as a result, it lessens the length of the active day in which animals can fulfil their other biological needs (foraging, social interaction, and so on.). The knowledge of the variables that impact this component of complete resting time may then enable us to create methodologies to safeguard biodiversity.

The differences in temperature, mean yearly temperature and the percentage of leaves in the daily intake of food impacts the enforced resting time as established by the assessment. The enforced resting time necessities are to a limited extent set by the time needed for digestion (principally forced by the procedures of fermentation of leaves).

The differences of temperature and temperature most impose their impacts both directly and indirectly. The thermoregulation is a phenomenon in animals (animals tend to rest more when temperatures are high) and this is the direct impact of temperature on animals (Stelzner 1988; Dunbar 1992b; Di Fiore and Rodman 2001). According to Ventura in 2005 the auto grooming and scratching (exercises that can only takes place while the animal is generally at rest) increase with temperature.

The quality of the vegetation gets impacted by the temperature indirectly (for example the digestibility and quality of leaf material: van Soest 1982; Dunbar 1992a; Zvereva and Kozlov 2006; Korstjens and Dunbar 2007). This implies folivores should devote more time to take rest simply to permit the fermentation of leaf material.

Initially, we demonstrated that less folivorous species have a bigger zone at which anticipated resting times are inside adequate levels (for example happen on the two continents or in the environmental adjustments of their particular primate frugivore species. The last appears the almost certain: The majority of the African frugivores are terrestrial or semi terrestrial, that permits them to dwell in high temperature less forested areas, more dry forest than those dwell by South American frugivores, and they all are arboreal.

The Old-World primates have learned and living the terrestrial lifestyle and it bring with it a degree of environmental flexibility that remained denied to their New World sister taxa. There is no doubt, the main exemptions to the general standard that forecasted enforced resting time is greater at regions where species are missing in comparison to where they are available.

There are two African genera Papio and Chlorocebus. These two genera are terrestrial open-country specialists and their distributions may be limited by other time budget components. The high-altitude locations are prime primate habitat as the linear relationship between normal yearly temperature and resting time as stated by experts.

Figure 7.7. Vervet Monkeys (Chlorocebus pygerythrus).

Source: Image by Flickr

These types of relationship should not hold as there are two reasons to not anticipate that. To begin with, areas at exceptionally high altitudes frequently have low oxygen levels and decreased vegetation cover, in this manner prompting environmentally increased demanding conditions (other than for a couple of extreme dietetic specialists, for example, the African gelada monkey, Theropithecus gelada).

Second, the positive connection among temperature and resting time may hold only inside the temperature range for which experts have behavioral data; in the very less temperature range (where we have without a doubt, not very many data points), the association might invert and resting times may turn out to be adversely associated with temperature on account of the expanded requirement for huddling and preservation of energy as well as oxygen (Scantlebury et al. 2006; Wada et al. 2007).

To illustrate the potential uses of the equations, one of the optimal ways is to assess and understand the relationship between enforced resting time and as well as climatic conditions to evaluate the impact of an increase in average annual temperatures on primate biogeographically distributions.

At overall increases in temperature of 2 degree Celsius and 4 degree Celsius, the spreading of habitat suitable for folivores will be significantly minimized. The dispersion of appropriate habitat would become increasingly fragmented and progressively restricted. This problem would cause significant impact on Old World folivores.

Although some variation exists from location to location, none the less the scale of the effect is enough great that the implications for the survival of many of the dietetically more specialized primates is worrying. The research and analysis validate consideration of resting time only. While resting time plays very important role in determination of presence/ absence, it is definitely not the only one.

In fact, both moving and feeding time are impacted by temperature or as well as variation in temperature in both folivorous and frugivorous primates (Dunbar 1992a, b; Korstjens et al. 2006; Korstjens & Dunbar 2007; Lehmann et al. 2007a, 2008a; Dunbar et al. 2009; Willems & Hill 2009), and the multiplied impact of these relationships will only help in minimizing further the range of habitats within which the primates can be easily survived.

In addition, our findings only helpful in partly incorporating the effect that temperature has on leaf quality which results in causing exponential effect of temperature on time budgets. Therefore, the long-term expectation, is that folivorous primates will gradually reached to level of extinction in the Old World (but possibly less so in the Americas) as global temperatures increases. It means that the future policies on conservation should primarily focuses on preserving the ever-decreasing habitats suitable for folivorous primates (such as the already endangered ones such as colobines and gorillas).

In conclusion, it was explained that enforced resting time is influenced primarily by temperature variation, temperature, as well as diet. An empirical equation is described in order to evaluate or assess the resting time for any primate based on these variables. The possibility that enforced resting time might cause very significant impact on the biogeography of a major mammal group that was previously being ignored.

This study signifies that it is worth to take time budgets into consideration when examining or evaluating the likely impact of the climate change on the overall animal distribution patterns in the past as well as in the future. One of the other important things to note in biodiversity is that temperature variance (i.e. seasonality) plays a very crucial role in predicting the ability of species to survive in the uncertain climatic conditions.

However, climate change models typically only helpful in predicting the rainfall and mean temperature. Finally, it is important to note that the importance of this approach over the more conventional bivariate conservation models that plot presence and absence against the variables of climate is that the present models helps us in evaluating the stress that a species can tolerate or possess the ability to tolerate throughout its range (Dunbar et al. 2009).

In contrast, bivariate models help up in explaining how the species' biogeographical boundary is likely to divert, and not how much it will have effect on the populations within the central parts of the range. In principle, it is entirely possible for a taxon's geographical range to be not affected at all by the climate change, but significantly reduction in its population size (e.g. Lehmann et al. 2008a). Thus, in effect, enforced resting time helps us in providing an index of the biological stress that a species faces in any given location.

7.5. MOLECULAR CLOCK STUDIES AND THE BIOGEOGRAPHY OF MADAGASCAR

Some experts have referred to the origin of Madagascar vertebrates as one of the most exciting enigmas that biogeographers have encountered. Simpson predicted that sweepstake dispersal events should take place at random intervals. However, this fact is not supported by mammal data.

Poux et al, estimated some molecular divergence dates for sweepstakes dispersal events. They might have occurred at an early point in history of clade and was never repeated again. Masters et al proposed a reconsideration of the molecular and geophysical data because it would support in dissertation of the apparent paradoxes. This data lays the foundation for the model described in this chapter.

Some authors like Yoder and Nowak presented a meticulous review of the molecular clock literature on Malagasy taxa. In all researches encompassing plants, the fossil-calibrated clocks revealed that Madagascar clades were not beyond 80 Ma. Hence all of them belong to Gondwana dispersal. It cannot be attributed to vicariance. All the researches showed the same conclusion, with just one exception.

Various molecular dating studies have been conducted on Malagasy invertebrates and all of them have supported the fact that dispersal was the cause. This is because the fossils indicate that most of the divergence happened post-Mesozoic.

There are however a couple of sequenced groups, namely, freshwater fishes and cichlids whose presence on Madagascar has been explained as a consequence of vicariance. Fossil calibration was not at a part of the molecular dating studies pertaining to this group. On the other hand, Madagascar-India-Sri Lanka and Madagascar-Africa-South America, the two major molecular clades had their vicariant distribution correlated with tectonics. The calibration used here was the opening of the Mozambique Channel.

Some of the experts like Yoder and Nowak came to the conclusion that dispersal had significant importance. However, the only thing the mentioned research indicates is how significant the calibration method is. In the recent researches on cichlids, fossil record is not used. Instead the splitting up of the Gondwana has been used to calibrate the molecular phylogenies. This has generated remarkable results.

It has been discussed that fossil-based molecular clocks can provide the maximum absolute ages by transmogrifying the minimum ages. The clock study of the Malagasy mammals is a proof of this fact. The authors discovered that the first ones to diverge were the lemurs and their African sister group. This was followed by the equivalent splits of African tenrecs, then carnivores and rodents.

In order to calibrate the clock calculations, six proven fossil dates were chosen. This was then transmogrified into maximum dates. The maximum and minimum dates were allocated to five fossil clades. The sixth one was allotted a minimum date.

It was estimated that the minimum age of the primate radiation was 63 Ma and the maximum was 90 Ma. However, this conclusion does not seem logical. Establishment of even a single maximum date for any taxon would be a simpler task. However, no maximum date was found in the mammalian fossil record.

In another study, Poux et al again employed transmogrified data for the purpose of inferring young dates pertaining the to the trans-Atlantic disjunctions in primates. This was the basis for accepting trans-Atlantic rafting.

Some experts arrived at the judgement that the various degree of differentiation which the Malagasy mammals and their sister groups diverged at different times. As a result, colonization happened.

The vicariance model, in contrast, various degrees of difference are an evidence of prior aspect of genome architecture. Some molecular clock analyses accept the degree of difference. These studies suggest that all the endemic that belong to Madagascar can be attributed to a different event. This is related to no other event. The community-wide vicariance has not contributed much towards it.

The same inference can be drawn for all biogeographic pattern. This is because, all patterns comprise of clades of different rank. In Madagascar, the vicariance in the orders of mammals indicate that their spatial differentiation is an outcome of having a common Earth history. The various degree of differentiation in all the four has no relation to time.

The primates are closely related to the rodents and the biogeography of this larger group which is about 40 % of all mammal species and so is relevant. There are trans – Atlantic disjunctions which have already been mentioned. Rodents have three primary clades and each one of them occur in America, Africa and Asia.

Despite this overlap, there are significant differences in the three clades, especially with respect to the Madagscar: Hystricomorpha (porcupines, the basal group): east to Borneo (strepsirrhines); not in Madagascar. Sciuromorpha (squirrels): east to Sulawesi (haplorhines); not in Madagascar. Myomorpha (rats): east to Australasia, along with a diverse clade in Madagascar.

Figure 7.8. Porcupines.

Source: Image by Flickr

There are a number of authors who have commented on the surprising absence of the first two groups, especially the widespread squirrels from Madagascar and Australasia and so, the same pattern occurs in haplorrhines. There is a difference in the distribution of the rodents and that of the primates. Despite this, the main phylogenetic or biogeographic breaks (nodes) are same: Mozambique Channel, the Banda / Molucca Sea, east of Sulawesi, Makassar Strait (Wallace's line) and Atlantic Ocean.

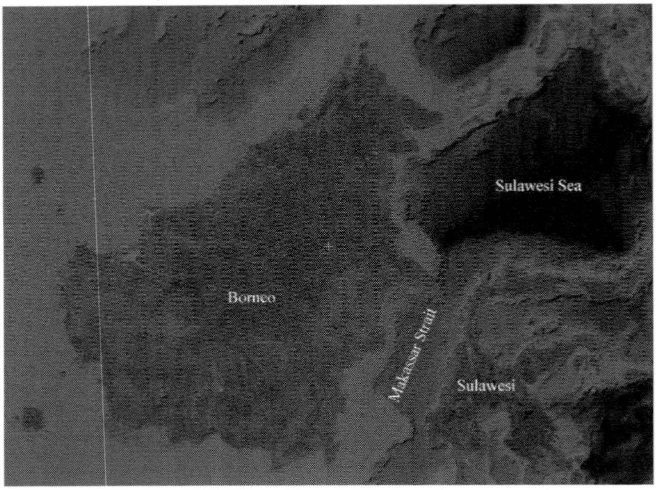

Figure 7.9. Makassar Strait.

Source: Image by Wikimedia

According to Horner, the rodents diverged from lagomorphs and that the three main rodent clades diverged from each other at around 60 Ma that is a minimum date along with all this evolution which are taking place in just 3.1 million years. A similar early as well as rapid differentiation is accepted for the main primate clades. The age which is proposed for the groups is much older than what it has been thought, even though the time involved in their differentiation can be much less.

REFERENCES

1. http://www.outrop.com/uploads/7/2/4/9/7249041/cheyne_09_challenges_and_opportunities_gibbon_rehabilitation.pdf [Accessed 27 Aug. 2019].
2. https://www.academia.edu/17604049/Resting_time_as_an_ecological_constraint_on_primate_biogeography [Accessed 27 Aug. 2019].
3. http://johngrehan.net/files/1813/3997/5296/Heads_2009_primates.pdf [Accessed 27 Aug. 2019].
4. https://www.ajol.info/index.php/ejesm/article/view/103875/93996 [Accessed 27 Aug. 2019].
5. https://www.researchgate.net/publication/226349591_Biogeography_and_Primates_A_Review [Accessed 27 Aug. 2019].
6. https://link.springer.com/article/10.1007/s10764-013-9664-1 [Accessed 27 Aug. 2019].
7. https://www.researchgate.net/publication/288995387_Conservation_Challenges_to_Preventing_Primate_Extinction [Accessed 27 Aug. 2019].

Chapter 8 Primate Progress and its Future

"Knowing that we are primates, I think, is a fascinating discovery, and a very interesting and rather cheering one"

– Christopher Hitchens

CONTENTS

Primates are the member of the biological order and so it is required to look up at their progress and future. The scientists who work on such things are referred to as primatologists. This chapter discusses about the primate evolution and future conservation. It also gives some insights about the future of homo sapiens and the optogenetics in primates and also the non – human primate optogenetics. This chapter explains the future of primate extinction and also about the primate conservation. There are some future directions regarding the primate conservation which are described in this chapter. In the end of this chapter, it provides information about the progress and prospects for genetic modification of non – human primate model in biomedical research.

8.1. INTRODUCTION

The primates are made up of 504 species in all and so, it is considered as the largest groups of mammals. There are some from these which we are familiar with as they are active during the day time but there are some others which do come out only in night. There are some of them who eat fruit, other eat leaves and there are some who eat insects and some hunt for meat. And there is also a difference in the food that they eat such as some are there who eat mushrooms and even there are those who eat other primates.

Figure 8.1. Primates.

Source: Image by Pixabay

Primates mostly live in tropical rainforests and around two third of these species can be easily found in Brazil, Madagascar, Indonesia or in the Democratic Republic of Congo. The primates can also be found in grassland, snowy mountains and deserts and an even thrive in the cities.

There are primates who can live in solitude, can live in huge as well as complex societies or anything which comes in between these two. Some live on trees and there are some who do not move much at all.

Primates are very necessary for tropical rainforests, pollinating trees and dispersing seeds across these vital carbon stores. They are the closest biological relatives, and we look to them to understand our own evolution.

8.2. PRIMATE EVOLUTION AND FUTURE CONSERVATION

During the past 15 million years, model of ape history has been shaped through the study of genetic variation in a large panel of humans, chimpanzees, gorillas and orangutans.

The most comprehensive catalog of the great ape genetic diversity demonstrates the evolution and population histories of the great apes from the Africa as well as Indonesia. The resources are also likely to help in the current and the future conservation efforts which strive in order to preserve the natural genetic diversity in the populations.

Figure 8.2. Orangutan great ape.

Source: Image by Flickr

A number of scientists and wildlife conservationists more than 75 assisted the genetic analysis of the 79 wild and captive – born great apes. The represent all the six great ape species such as chimpanzee, bonobo, Sumatran orangutan, Bornean orangutan, eastern gorilla and western lowland gorilla and the seven sub-species. In the sampling, the nine human genomes were also included.

There are a number of various genetic variations among the great apes which had been largely unfamiliar, due to the difficulty in obtaining the genetic specimens from the wild apes. In many countries, some of the conservationists have helped in recent effort and researches.

A UW graduate student in genome sciences, Sudmant said, "Gathering this data is critical to understanding differences between great ape species and separating aspects of the genetic code that distinguish humans from other primates." The analysis done on the great ape genetic diversity is likely to reveal the ways of the natural selection, population growth and collapse, geographic isolation and migration, climate and geological changes and other factors shaped primate evolution.

According to Sudmant, the learning more about the great ape genetic diversity helps in contributing in the knowledge about the disease susceptibility among the variety of primate species. Such kind of questions are important for both the conservation efforts as well as to the human health. For the deaths of a large number of gorilla and chimpanzee in Africa, the ebola virus is responsible and it is also responsible for the origin of HIV, the virus which helps in aiding the AIDs is SIV that is Simian Immunodeficiency Virus. Sudmant and Eichler wrote in a companion paper that they inadvertently found the first genetic evidence in a chimpanzee of a disorder resembling the Smith – Magennis syndrome, disabling the physical, mental and behavioral condition in humans.

While the researchers were exploring and comparing the accumulation of the copy number of variants, the discovery came out during the great ape evolution. The genomes of the human and great apes have been re – structured by the duplication and deletion of DNA segments and are behind many genetic diseases.

Along with offering the view of the origins of humans and their disorders, the new resource of ape genetic will help in addressing the challenging plight of great ape species on the brink extinction. The resources are said to be the important tool in order to enable the biologists so as to identify the origin of great apes poached for their body parts or hunted down for bush meat.

The research also provides explanation about why the current zoo breeding programs, that have been implemented in order to increase the genetic diversity of the captive great ape populations, have led to captive ape population which have genetically dissimilar to their wild counterparts.

Sudmant said, "By avoiding inbreeding to produce a diverse population, zoos and conservation groups may be entirely eroding genetic signals specific to certain populations in specific geographic locations in the wild." By the researchers, one of the captives – breed apes' study has been done which had the genetic makeup of the two distinct chimpanzee subspecies which are located more than 2000 kms away from each other.

The research also describes many changes which occurred along each of the ape lineages as they are separated from each other through migration, geological change and climate events. There is formation of rivers, the partition of islands from the mainland and other natural disturbances which have all served in order to isolate the groups of the apes. The isolated populations can then be exposed to a unique set of environmental pressures, that lead to population fluctuations and adaptations which depends on the circumstances.

Despite the fact that early human like species were present at the same time as the ancestors of some present-day great apes. Furthermore, the researchers found that the evolutionary history of ancestral great ape populations was far more complex in comparison to that of the humans.

Figure 8.3. Human are closest relatives of Chimpanzees.

Sources: Image by Pixabay

While comparing the humans with their closest relatives that is chimpanzees, human history appears to be boring according to Sudmant and his mentor Evan Eicher. From last few million years, the evolutionary history of the chimpanzee is fraught with the population explosion and then the implosions demonstrating the remarkable plasticity. However, the reasons for the fluctuations in chimpanzee population size long before our own population explosion ae still not known.

Sudmant stated that he has interests in studying the great apes and he is interested in preserving the great ape's species and these interests stems from the similarity of the great apes to humans and their curiosity about us. He said, "If you look at a chimpanzee or a gorilla, those guys will look right back at you. They act just like us. We need to find ways to protect these precious species from extinction."

8.3. WHAT MAY BECOME OF HOMO SAPIENS?

While comparing our use of genetic manipulation, it is even less possible to predict the manipulation of machines or machine manipulating us. There are a number of writers who have predicted that there can be a link in between our bodies with the robots or upload our minds into computers. In fact, we the humans are already dependent on the machines. As much as the humans build them in order to meet the human needs, the humans have structured their own lives and behavior so as to meet theirs.

Machines are becoming more complex and interconnected and with this the humans will be forced to try to accommodate them. This view was conveyed by the George Dyson in his book named as *Darwin among the Machines* which was published in 1998. He wrote, "Everything that human beings are doing to make it easier to operate computer networks is at the same time, but for different reasons, making it easier for computer networks to operate human beings. Darwinian evolution, in one of those paradoxes with which life abounds, may be a victim of its own success, unable to keep up with non-Darwinian processes that it has spawned."

Figure 8.4. Charles Darwin who gave the Darwinian Theory of evolution.

Source: Image by Maxpixel

The technological ability of the humans threatens to swamp the old ways which evolution works. An essay in 2004, by the evolutionary philosopher Nick Bostrom of the University of Oxford, he considered the two different views of the future.

One was the optimistic side in which he wrote, "The big picture shows an overarching trend towards increasing levels of complexity, knowledge, consciousness, and coordinated goal-directed organization, a trend which, not to put too fine a point on it, we may label 'progress.' What we shall call the Panglossian view maintains that this past record of success gives us good grounds for thinking that evolution (whether biological, memetic or technological) will continue to lead in desirable directions."

Even though the reference to 'progress' surely causes the late evolutionary biologists Steven Jay Gould to disfavor the point that can be made. The arguments made by Gould was that the fossils including those from our own ancestors, tell us that the evolutionary change is not a continuous thing, instead of this, it occurs in fits and starts, and it is also not progressive or directional.

Evolution has indeed led to at least one vector that is towards the direction of increasing complexity. It can be said that it is the fate of the future human evolution and increased the complexity through some combination of anatomy, physiology or behavior.

If the adaptation is continued by the human, there is no genetic or evolutionary reason that the humans cannot still be around to see the sun die. Extinction is not similar to aging and therefore, the extinction does not appear to be genetically programmed into the species.

Bostrom provided a vision of how the uploading of the human brains into the computers can lead to failure. The advanced artificial intelligence technology can include the various components of human cognition and thus, get back together those components into some other thing which is not a human and that will make us outdated.

As per the prediction of Bostrom about the course of events, he stated that, "Some human individuals upload and make many copies of themselves. Meanwhile there is gradual progress in neuroscience and artificial intelligence, and eventually it becomes possible to isolate individual cognitive modules and connect them up to modules from other uploaded minds.

Modules that conform to a common standard would be better able to communicate and cooperate with other modules and would therefore be economically more productive, creating a pressure for standardization. There might be no niche for mental architectures of a human kind."

Bostrom concluded with an even more dreary possibility as if the technological ancientness were not disturbing enough. He stated that if the machine efficiency became the new measure of the evolutionary fitness, human will be discarded of the family.

As per the statement written by the Bostrom, he wrote, "The extravagancies and fun that arguably give human life much of its meaning—humor, love, game-playing, art, sex, dancing, social conversation, philosophy, literature, scientific discovery, food and drink, friendship, parenting, sport—we have preferences and capabilities that make us engage in such activities, and these predispositions were adaptive in our species' evolutionary past; but what ground do we have for being confident that these or similar activities will continue to be adaptive in the future? Perhaps what will maximize fitness in the future will be nothing but nonstop high-intensity drudgery, work of a drab and repetitive nature, aimed at improving the eighth decimal of some economic output measure."

"The extravagancies and fun that arguably give human life much of its meaning—humor, love, game-playing, art, sex, dancing, social conversation, philosophy, literature, scientific discovery, food and drink, friendship, parenting, sport—we have preferences and capabilities that make us engage in such activities, and these predispositions were adaptive in our species' evolutionary past; but what ground do we have for being confident that these or similar activities will continue to be adaptive in the future? Perhaps what will maximize fitness in the future will be nothing but nonstop high-intensity drudgery, work of a drab and repetitive nature, aimed at improving the eighth decimal of some economic output measure."

8.4. OPTOGENETICS

8.4.1. Optogenetics in Primates

The techniques which is used for controlling the sub populations of neurons in the complete brain using light is known as optogenetics. The technique contains the ability to enhance the basic systems neuroscience research and to inform the mechanisms and treatment of the brain injury and the disease. Prior to launching the large – scale primates' studies, the methods are required to be further categorized and also adapted to be used in the primate brains.

The assessment has been done of the safety and efficiency of two viral vector systems (lentivirus and adeno-associated virus), two human promoters (human synapsin (*hSyn*) and human thymocyte-1 (*hThy-1*)) and three excitatory and inhibitory mammalian codon-optimized opsins (channelrhodopsin-2, enhanced *Natronomonas pharaonis* halorhodopsin and the step-function opsin), which we characterized electrophysiologically, histologically and behaviorally in rhesus monkeys (*Macaca mulatta*).

A new device for measuring in vivo fluorescence over time have also been introduced which allows minimally invasive assessment of construct expression in the intact brain. There are a set of optogenetic tools which is designed for the optogenetic experiments in the non – human primate brain.

8.4.2. Non - human Primate Optogenetics

Optogenetics is a technique which is the use of genetically coded, light – gated ion channels or pumps (opsins) for controlling the millisecond resolution of the neural activity. Optogenetics can expand the human understanding of the neural basis of the normal and pathological behavior and this can be done by targeting the opsin expression to the particular cell types and neuronal pathways.

Optogenetics is required to be applied in the study of non – human primates, in order to maximize the potential of optogenetics so that the study of the human cognition and behavior can be done. The homology in between the NHPs that is the non – human primates and humans make these animals the best model for experiments so as to understand the human brain function and dysfunction. In order to have the traditional promise by the genetic tools, their use must be demonstrated effectively in large, wild animals like the Rhesus macaques.

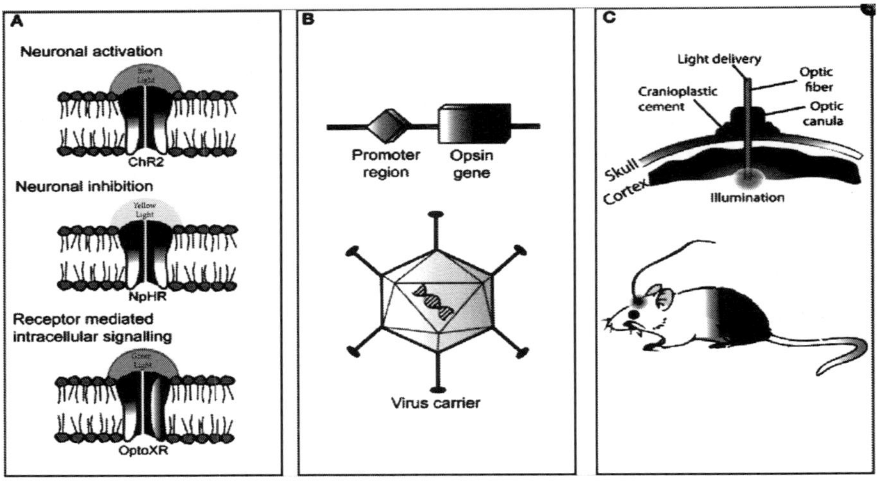

Figure 8.5. Optogenetics Stimulation.

Source: Image by Wikimedia

The human perception, cognition and action are majorly dependent on the specialized brain networks which arise during the evolution of primates. This specialization provides non – human primates, especially Old-World monkeys like the Rhesus macaques which is an invaluable animal model which can investigate the human brain function.

The modern neuroscience has been reshaped and this is done by the technologies which are proved to be game changing including the optogenetics and genetically coded calcium indicators which are most effectively applied on to the transgenic animals. Even though the transgenic Rhesus macaques have been created successfully, the economic considerations and the long generation time can restrict the utility of the transgenic Old-World monkeys in neuroscientific research.

Above all, the successful implementation of emerging technologies such as the optogenetic in wildtype macaques will provide a rich foundation for the translational application of circuit-based therapy to the humans. Therefore, the developing methods, which are used to facilitate the application of the genetic techniques so as to control and monitor the neural activity in the non – human primates, is one of the most critical challenges in the modern neuroscience.

Recent studies have shown the effective application of the optogenetics to the non – human primates and this is done by demonstrating cell type particular expressions by using the gene promoters, improved light delivery, selective projection targeting and opsins with improved light sensitivity. Cell type specific promoters have been used in order to study the function of konio cells in the LGN, dopamine neurons in the midbrain and Purkinje cells in the cerebellum.

The non – human primate neuroscience is brought one step closer to the capabilities of the transgenic mouse studies by using the cell type specific promoters. Furthermore, the development or discovery of new regulatory sequences that give selective expression in a wide array of neuronal types and this will be one of the greatest challenges which we can face. There are a number of genes which are expressed in many cell types instead of being particular about the cell type.

Along with this, this is not clear that how much conserved the non – coding transcriptional regulatory regions are and therefore, it is also not clear that which mouse promoters will be proved to be effective in the monkeys. For the advancement in the field, studies can hold the major key to provide a wide array of cell type specific promoters and these are the studies in which screening of the promoters can be done in a high – throughput fashion.

8.5. FUTURE OF PRIMATE EXTINCTION

Extinction is sweeping the flora as well as fauna of the planet. The conservation biologists have this in their knowledge that it will not be possible to save every species from extinction, given the speed of the process.

The main problem is that only a part of the animals and plants on the earth have been named and identified. And this leading to extinction of a number of creatures and still it is going on even before anyone get to know about their existence. And therefore, this has been an interesting thing for the researchers to find shortcuts for this problem by using well known, highly visible groups of animals like the birds or the butterflies as 'proxies' for the biological diversity.

The technique is that the number of the indicator species such as birds and butterflies in any area, will be reliable guide for the total number of all the species. There is an area which is rich in indicators species will be regarded as a much worthy candidate for conservation than one which is poorer in species.

Extinction is something which means more than the disappearance of species. There is a contribution done by each species in the health and stability of the overall ecosystem, and so some species will be missed more than others.

Jukka Jernvall of the University of Helsinki, Finland, and Patricia C. Wright of the State University of New York, Stony Brook, examine the problem in a report in *Proceedings of the National Academy of Sciences*. For the role of test case, they use primates like monkeys, lemurs, apes and so on. They are animals and so as primates are highly visible. They use animals because the zoologists know all the habits well. And many of them are also threatened with extinction.

These kinds of threats have been ranked by the conservationists, and therefore, researchers can get an idea of the order in which the primates are likely to become extinct. Those species which are listed by the International Union for the Conservation of Nature (IUCN) for example, as 'critically endangered' then they are regarded as more likely to become extinct than species in the category 'vulnerable'. These species are more likely to be excluded from the Earth than the animals which are referred to as abundant or not threatened.

The researchers predicted the future in which the critically endangered primates disappear, followed by those which are listed as merely 'vulnerable'. Critically, they work out on the way that how these waves of extinction will influence the ecological balance of primate communities, provided that each species of the primate cleaves to its own habits like some wat fruits, others insects, some prefer hanging around the trees, others just live in more open area and many more.

Jernvall and Wright also categorized their scenario by regions. By doing so, they shown the effects of these waves of future extinction that can vary depending upon where they are going to occur, simply quantifying the extinction by listing the species without any reference to their habits and thus, provides a misleading picture of the effects of extinction on ecology.

Let's take an example, the ecological effects of the primate extinctions are least severe in South America despite the fact that the loss of the species there will be great. This is the reason why the primates are all members and the group are named as the 'New World' monkeys.

The ones which are in the South America tend to have same habits while considering against the diversity of primates as a whole, therefore, the ecological roles of species lost will be taken over by other which are similar species. As per the thinking of the researchers, South America can lose more than half of its primates by applying a little change in their ecological composition.

In Africa and Asia, the ecological effects of the extinction will have greater effects at the place where the diversity is higher. This is because then the monkey will not be the only one who will have to suffer. The great apes which have large bodies and have instead different ecologies can also disappear.

The major effects can be seen on the island of the Madagascar which is considered as the home to the distinctive primates known as lemurs. Unlike anywhere else in the world, they have diversified on this island, and thus, fills an enormous range of the ecological niches that in some other places would be occupied by other animals, not specifically the primates. In comparison to 8 to 12 % anywhere else in the world, in Madagascar, primates make up 44 % of the terrestrial mammal species excluding bats.

But there can be one more reason for why the effects of extinction will be disproportionate in Madagascar. Similar to all the islands with specialized, unique faunas, Madagascar has suffered more than large, continental land masses from extinction.

Figure 8.6. Archaeolemur Edwardsi which are no more found in Madagascar.

Source: Image by Wikimedia

There are a number of species of lemur which have already become extinct, since the very first time the human being colonized the island about a thousand years ago. Some of these lemurs were large, and so occupying niches which would be associated with the apes that are still living in the Africa and Asia.

This provides with an impression that the extinction will have a more severe effect on the ecology on the Madagascar than anywhere else in the world. But according to the researchers, it can be what they are seeing is the final decimation of the relevant primate habitats in the Madagascar. Thus, the Madagascar primate fauna can be one step ahead of everywhere else in the extinction stakes.

The bad news is that the primates are in dire trouble. Approximately 60% of the species are in threat of extinction and 75% are declining in numbers. If no suitable actions are taken, then these numbers will grow, and more species will disappear forever.

The humans are the ones who are responsible for such an awful situation. Primates are majorly threatened by the reason of losing their habitat when it is logged or converted into farms or ranches. They can also be hunted for meat or in order to supply the illegal trade in pets and the body parts. In the list of threats, the addition can be of the road construction, oil and gas extraction, mining, pollution, disease and climate change.

People who are residing in these areas do not try to destroy the primate species. Researched have talked with the people who are driven to the distraction when their subsistence crops are ruined by the wild animals that also includes the primates, with the hunters who hunt in order to feed their family and also to earn money for their expenses and with the fisherman who hunt primates in order to feed their family in case when no fishes are available.

Figure 8.7. Lemur.

Source: Image by Maxpixel

The threat to primates has led to the political uncertainty, organized crime, socio – economic instability, corruption and policies which are in support of the short – term profits over long – term sustainability.

There must be conservation policies which should be adopted by each country, habitat or species on the basis of the exact nature of each and every problem. This is an alarming task. But this kind of hope that people develop all stems from the exceptional projects and extraordinary people who are present around the world such as Virunga rangers who risk their lives in order to protect gorillas and other primates in the DR Congo or Fundacion Pro-Conservation primates Panamenos which is the organization dedicated towards conservation of Panama's monkeys.

The solutions lie in the global fairness and further taking the responsibilities for the implementation of the actions of the human. The political and human disasters in other countries cannot be ignored. It is required to accommodate the needs of both primates as well as people in land use policies, monitor primate populations and mitigate illegal trade. In order to rescue the species from the brink of extinction, captive populations can be used, but in case if their natural habitat is gone then there is no hope.

In order to reduce the pressure on the primates and their habitats, it is required to decrease the demand for tropical hardwood, beef, palm oil, soy, rubber, minerals and fossil fuels and further promote sustainable resources. It is required to make it unacceptable to prioritize excessive and unnecessary consumption over the persistence of other species. Every consumer decision made by the human has global implications. Furthermore, it should be understood that the wild animals are not suitable to be kept as pets.

8.6. PRIMATE CONSERVATION

From the species and the sub species of the primates which are approximately 600 that are living today, around half of them are in danger of extinction. This is an estimate which has increased alarmingly over the preceding decade. In fact, one subspecies in West Africa, Mis Waldron's red colobus (Procolobus Badius Waldronae), can already be extinct.

Moreover, the number of serious threats which have been recognized to primate survival has increased over the last decade. Diseases were recognized as the important area of study a decade ago, and so they were not considered as a factor which can threaten primate populations with the local extinction.

Contrastingly, in today's era, there are a number of well documented cases of the dramatic primate population decreases which is caused by diseases and so, it is considered as an important factor in conservation planning, specifically for those populations which are in danger of contracting the diseases from the exposure to the humans. In the similar way, in the last decade, climate change was the major thing which was largely ignored by a number of countries to a phenomenon of grave concern.

It is obvious that a number of primates are endangered by the actions of the humans and it is important to place great efforts in conserving the forests that they inhabit. However, it is believed that there is a reason to be optimistic because the rate of the forests loss is declining, a great deal is now known about the primate diseases and the programs in order to decrease the bushmeat industry which are being successfully implemented.

This optimism should not force the humans to reduce the efforts, instead this knowledge of the threats provides a great opportunity for scientists so that they can make significant contributions to the primate conservation. The researches which were done over the last few decades have laid out very clear objectives which can be met in order to promote the primate conservation.

An example can be by offering people sustainable alternatives to the bushmeat trade, decreasing the deforestation and promoting the reforestation, making protocols and developing parks in order to reduce the risk of the disease transmission in between the people and the primates and further to reduce the carbon emissions so as to curb the climate change.

Certainly, there are a number of questions which needs to be considered in order to make primate conservation a reality. These include:

- What will direct depopulated areas into a process of ecological recovery or into plantation economies?
- How can conservation gains be made from declining rural populations in a socially responsible way?
- What are effective measures to reduce the transmission of diseases among people and non-human primates?
- How is it possible to reduce the national and international trade in bushmeat?
- How will climate change impact specific regions and primate species?

All of these questions provide the foundation upon which to build research programs that will make a difference and provide the information needed to construct informed conservation/management plans.

8.7. FUTURE DIRECTIONS FOR PRIMATE CONSERVATION RESEARCH

The researchers who are researching on the primates have made many positive contributions that includes helping to form the national parks, founding or leading organizations in order to promote the awareness and fund conservation activities such as Jane Goodall's Roots and Shoots, Russell Mittermeier at Conservation International, raising the international profile of the primates which are threatened and contributing to the successful management of the severely endangered taxa such as mountain gorillas.

Figure 8.8. Jane Goodall in LA on Roots & Shoots.

Source: Image by Flickr

At the field stations, the presence of the primate researchers can facilitate the law enforcement, further provides alternative income to local communities and also promotes awareness of the importance of biodiversity and its protection. The students getting involved in the research projects can help in training the next generation of the conservationists and the natural resource managers.

In a number of habitat countries, such kind of training is specifically important where the capacity building is vital to ongoing conservation efforts and mostly an important activity at the primate field sites. Primate field researches shows the status of the threatened species, assesses the effects of the threats, examines interactions in between the non – human primates and the human primates and further demonstrates the major ecological role that various primates play in the ecosystems.

The much basic primate research is done considering the primate conservation. Even the studies which are not explicitly geared towards conservation such as those which focuses on the social behavior, can never have a positive impact by raising interest in primates in the broader public.

8.7.1. Fill gaps in taxonomic and geographic knowledge

In comparison to other tropical taxa, the primates are better studied. Despite this fact, human lack even the most basic information about the distribution, status of the population and threats for many primate taxa. The field studies which are published on the primates are not randomly distributed across the phylogenetic space.

Let's take an example. The recent work, which was published in the International Journal of Primatology, is excessively skewed in the direction of apes and away from the strepsirrhines. There was a systematic examination done on the information which is available on the web about each primate taxon and that examination showed a heavy bias towards the apes and also, showed that the information which is available publicly on primates did not correlate with extinction risk.

This shows that for some taxa, the basic field research or population surveys will make a real contribution. And if there is no field research, the human can be ignorant of the fundamental taxonomic units of conservation, and so it becomes difficult to target the actions taken for conservation towards a taxon that the humans do not know exists.

In the similar way, without having a good knowledge about a species' distribution and density, it becomes difficult to construct meaningful management plans. Even if one eschews species as the fundamental units of conservation, additional field research and surveys are able to provide valuable information about the distribution of threats as well as identify geographic areas or primate communities in need of conservation attention.

8.7.2. Make behavioural research more relevant to conservation

The behavioral research on the primates can make a major contribution to the conservation that it is doing at present. In order to date much behavioral work on primates has had limited relevance to applied conservation.

The researches have been done which focuses on what are the typically referred to as the ecological topics such as feeding ecology, seed dispersal, demography and population density, and it is often more obviously relevant to conservation than research on the topics which are classified as behavioral like the social interactions, mating behavior, endocrinology, locomotion, communication, knowledge of individual behavior which is nevertheless indispensable.

There can be contribution done by the behavioral studies in order to conserve and manage by documenting the behavioral plasticity of species, examining their interactions with the humans, further informing the reintroduction and captive breeding programmes, characterizing dispersal and anticipating responses to habitat destruction and climate change.

There are a number of good examples of primate behavioral research with conservation applicability, which includes the work on crop – raiding, dispersal in fragmented landscapes, disease transmission and behavioral responses to hunting and logging, this kind of work are needed more.

8.7.3. Increase research in marginal habitats and outside protected areas

There is a requirement of more research on primates which are living outside the protected areas, inhabiting marginal and degraded habitats and ranging across complex, multi – use landscapes. Since, the primate's habitats all around the world, they are being fragmented and degraded and therefore much of the future primate conservation will take place considering the suboptimal habitats and environments which are quite different from those to which a number of species are best adapted.

Although, this point is gaining appreciation which is increasing and a great attention is being paid to examination of the primates which are living in the suboptimal ecological conditions, there are some other research which are still being conducted inside the protected areas and at sites which comprises relatively high - quality habitats and which are relatively undisturbed.

For example, a recent survey was done of the research which was published from all the great ape range countries in Africa and Asia and such research directed the attention towards the protected areas which are strongly biased towards the large national parks which contains great apes. This kind of bias can provide dangerously optimistic impressions of population vital rates and further result in the crucial gaps in our knowledge which are of whether and how the primates survive, adapt and reproduce in lower quality and degraded environments.

In such cases in which the majority of the individuals live outside the formally protected areas, it can also mean that the research results can proved to be the representative of the minority of the individuals of species which the human seek to protect.

The researches can be rendered by the major focus on the primates which are outside the protected areas, in disturbed and degraded habitats and also occupying such mosaic habitats which comprises the mixture of the land covers and uses and can make the researches can directly be applicable to the conservation and management of threatened primate populations.

8.7.4. Expand climate change research

The climate change research includes the heavy use of species distribution modelling and climate projections, should also incorporate the biotic interactions in an effective manner like the competition and resource availability and the dispersal abilities of primates.

A number of studies on the effects of climate and land – cover change on primates and other mammals model habitat suitability are basically based on the ecological variables like the temperatures, rainfall, altitude or fairly course – grained indices of the land cover or a combination of these variables along with the time budgets.

Despite the fact that these variables are clearly important and also can predict the broad patterns of distribution and density, they are likely to be less successful on the local scale where the differences in presence – absence and population density occur in between the areas which have much similar values for basic ecological variables.

Such kind of small – scale cases measures the disturbance or food availability which are normally better predictors. Furthermore, the correlation in between the basic climatic variables such as temperature, altitude and rainfall and the factors which directly influence the population density and the distribution of the most primate species for example, the food availability, is more likely to become weaker as the global climate changes.

The reason can be for example the plant food distribution that can change at a different rate in comparison to the basic ecological variables which influence it. For instance, the population density of white – bearded gibbons,

Hylobates albibarbis, at Gunung Palung National Park, West Kalimantan, Indonesia, decreases the predictably with the altitude, which is mostly due to the decline in the availability of figs, that are important fallback foods at the time of resource scarcity.

Figure 8.9. Hylobates albibarbis.

Source: Image by Wikimedia

The assumption can be dangerous that the warming of the higher elevation forests to the conditions which are superficially comparable to present day lowland forest types would mean that the higher elevation forests will immediately provide high quality habitat for gibbons. The reason can be the hemi – epiphytic figs and their host trees which are most important to gibbons that can take hundreds of years to establish and grow at the high elevations.

This shows that the more rapidly the climate changes, the les good simple ecological variable will be predicting the distribution of primate species over the time of the several decades at which the most local conservation decisions are made. There are field research and modelling which can predict the rates and directions of the changes in underlying ecological variables like the distribution and density of the key lowland food plants, in response to the climate change are greatly needed.

Many attempts have been made in the models in order to shift the species distribution which leads to the climate change tacitly and further assume that the species will become able to move in order to keep pace with the climates which is constantly changing. This kind of assumption is not likely to be true in case of primates which appear to be surprisingly restricted in their dispersal abilities.

At the same time, there is a recent model of mammalian dispersal abilities which suggests that the primates will be one of the mammalian taxa which is least able to move in order to track the changes in the climate. This shows that the studies of the primate responses to climate change should include species – specific dispersal abilities and that the projection which fail to do so will provide dangerously optimistic estimates of their abilities in order to adapt the climate change.

8.7.5. Promote recognition of the value of ecosystem services provided by primates

There are a number of primates which are present across the tropics and they all perform various ecological functions which are important for maintaining the healthy, well – functioning ecosystems for example pollination, seed dispersal, seed predation, folivory. There are a number of evidences that proves that the primates can be the sole providers of some of the ecosystem services and that the loss of primates can degrade ecosystem structure and function.

Although, they are very important for various ecosystems, the economic value of the ecological services provided by the primates is rarely estimated or implemented into the discussions of the policies in context to conservation and management. The important services provided by the primates are widely recognized and explicit consideration is done of their economic value can provide useful justification for the conservation of primates.

Immense contribution can be made by the primatologists to this endeavor by conducting the research which documents the ecological services which are provided by various studies and further promoting awareness of the crucial role which the primates play in order to maintain ecosystem function.

Researchers are required to work in order to provide the concrete information which will facilitate the process of setting priorities so as to determine the way of primate conservation funds that should be allocated. Since the funds and the attention are limited, trade – offs are inherent to conservation decision making.

Despite the fact that primates attract the substantial public interest and conservation funding, resources are not likely to be relevant in order to invest in the actions which will save all the primate populations and probably not even all the species from the extinction. This means that the decisions that are to be made regarding wise allocation of funds towards the particular conservation action which prioritize some populations, sites or species over others.

A key role is played by the primate researchers in this endeavor by providing the quantitative information which is necessary in order to conduct the prioritizations and by contributing their expertise to more subjective discussions.

There are two components which are involved in the funding conservation such as:

- The societal and political decisions about how much funding to provide, and
- The optimization decision about how to wisely allocate the available funds

Contributions can be made by the primate conservationists in both the areas by advocating legislation and increased funding and further engaging in decision – making processes which inform the allocation of conservation investment. In the case of the latter, primate conservation priority setting has to date generally which has been used in order to either identify the important locations for the conservation of specific taxa or to direct the attention towards the specific species or the populations.

There are future recommendations which are likely to be most useful for the conservation managers in case if they are framed in the form of concrete suggestions for some of the particular actions in specific places over the specific time frames and these available options are ranked in the transparent ways which requires costs, benefits, risks and uncertainty.

8.7.6. Embrace interdisciplinarity

The researchers are required to answer some of the most important questions related to the primate conservation which does not fit comfortably into the traditional academic categories. Similar to other realms of the conservation, much research which is required is interdisciplinary, necessitating close collaboration among a number of various fields in the natural as well as social sciences.

Primatologists are those who are often well versed in biology and ecology but also affiliated with the social science departments. These primatologists are well placed to spearhead such collaborations. The interdisciplinary work can be challenging. In that the things as simple as differences in terminology, dissimilar assumptions and distinct perceptions about the things which constitutes evidence that can complicate collaboration across disciplines.

Moreover, there is now a substantial literature which further provides concrete suggestions about the way to overcome such challenges and an increasing number of the examples of the successful interdisciplinary collaborations which further demonstrates a number of teams that have effectively negotiated them. Once frequently, other concerns expressed in relation to the interdisciplinary research and it seem to be diminishing.

For instance, there were new journals which were dedicated to the interdisciplinary research and was an increased appreciation of its importance by many editors and have alleviated reservations about the restricted outlets in order to publish the research that spans traditional fields.

There are new initiatives taking place at the universities, foundations and government agencies which fund collaboration across disciplines and they have eased the concerns regarding the limited financial resources for the interdisciplinary work. The widespread acknowledgement of its crucial importance for conservation, and further new incentives which promote it, suggests that the primatologists can increase their contributions in the conservation by embracing interdisciplinary research.

8.7.7. Acknowledge the value of applied work

Conservationists have long made the fact that there are some academic institutions which do not value applied conservation research as much as the theoretical work and fear this dissuades graduate students who aspire to the academic positions or junior faculty in order to seek tenure from the pursuing such research.

In addition to this, they have noted that the applied conservation research also often falls outside the traditional sources of funding, making pursuit of such difficult works, especially for the individuals at the beginning of their careers.

In recent years, this has shown considerable changes. Universities now – a – days value conservation work, there are a number of new graduate programs which are dedicated to the primate conservation that have emerged and there are a number of various contexts for conducting the conservation research outside academia.

In today's scenario, new funding sources for the purpose of conservation research appear every year and the primatologists who add conservation component to their research can well improve their chances of securing the funding. There is an increment in the appreciation of the importance of the applied conservation, the need for it grows ever more urgent and there are more opportunities than ever before in order to make the positive contributions.

All of these things happening in today's scenario are likely to make applied conservation work a more attractive option for the early career researchers. It is hope of many people and it is important to encourage the students to pursue such kind of work and their advisors in order support it.

8.8. PROGRESS AND PROSPECTS FOR GENETIC MODIFICATION OF NON - HUMAN PRIMATE MODELS IN BIOMEDICAL RESEARCH

The non - human primates such as the rhesus macaque has close relation with the humans and so this has directly led to the growing interests of the modeling human diseases using genetically modified non - human primates that is NHPs. The developmental paths of the NHPs are similar to that of humans in their anatomy, genetics, physiology and neural functions and their cognition, emotion and the social behavior.

The Non – Human Primates model have played an important role in the biomedical research in the development of vaccines and have assisted the reproductive technologies and new therapies for many diseases. Biomedical research has not been the primary role for the non – human primates. They have majorly been used for the safety evaluation and pharmacokinetics studies instead of determining the therapeutic efficacy.

The role of non – human primates' models have been revolutionized by the development of the first transgenic rhesus macaque in biomedicine. The development of the transgenic non – human primates' model of the Huntington's disease having some distinctive features, further suggested the uniqueness of the model system and the important role of the non – human primates' model for the human genetic disorders.

The modeling of the human genetic diseases using the non – human primates will continue to thrive and the reason for the same will be the latest advances in the molecular, genetic and embryo technologies. Non – human primates have a rising role in the biomedical research, particularly in the pre – clinical studies which is predictable. The path in the direction of the development of the transgenic non – human primates and the prospect of the transgenic non – human primates in their new role that will be in the future biomedicine, is required to be reviewed.

The advancement of the transgenic non – human primate has been driven by the availability of the high efficiency genetic tools.

In the United States, there is a commitment of the National Institutes of Health to the National Primate Research Centers. The recent development that took place in the Japanese consortium of the marmoset model for the neuroscience research, and the development of the primate research in the China and other nations, they both shows the importance of the transgenic non – human primate model. It is expected that the transgenic non – human model will become an important model system, particularly in the translational and preclinical studies along with the basic research.

The advancement of a transgenic non – human primate model assembles with the interest which is increasing in the traditional medicine which is further indicated by the establishment of the National Center for Advancing Translational Sciences. The animal model is very much important, and this allows for evaluating pharmacokinetic and toxicity and also for determining the efficacy of the novel therapeutic approaches.

Along with the criteria in order to select an appropriate model system which can maximize the outcome, the awareness of the importance of the longitudinal study has also increased in the field of biomedical both in research as well as clinical studies.

Moreover, there is no perfect animal model like drosophila, zebrafish, C – elegans, rodents, pigs and non – human models these are all unique model systems which can direct towards the unique questions which can further help in the advancement of the biomedicine. It is important to match the particular research questions with the best as well as relevant animal model.

Along with the awareness of the importance of the longitudinal studies which have increased with the time and has been shown by various latest reports on the longitudinal studies, the animal models such as the non – human primates simultaneously allow the longitudinal clinical assessment by using a number of clinical measurements which are a crucial component in the translational medicine.

In the upcoming future, the role of the transgenic non – human primate model of the human diseases in the preclinical study is expected to increase. The advancement will take place in the molecular tools like the ZFN and TALENs and this will further open new doors in order to model a wide spectrum of the human genetic diseases without limitation to the dominant genetic disorders.

Along with the predictable impact of the transgenic non – human primate model in the preclinical study, the expectation is that the interest in the basic research will increase. Non – human primate is an important model species in order to understand the neural computation, cognition and behavior.

In the coming future, the non – human primates will also be used for probing the circuit – level basis of the human neurological and psychiatric disorders, because of the greater similarity in between the non – human primate brain and the human brain.

REFERENCES

1. https://www.scientificamerican.com/article/what-may-become-of-homo-sapiens/ [Accessed 27 Aug. 2019].
2. https://www.ncbi.nlm.nih.gov/pmc/articles/PMC3814401/ [Accessed 27 Aug. 2019].
3. https://www.nature.com/scitable/knowledge/library/primate-conservation-is-the-cup-half-empty-53872551/ [Accessed 27 Aug. 2019].
4. https://www.nature.com/articles/nn.2749 [Accessed 27 Aug. 2019].
5. https://earthsky.org/science-wire/primate-evolution-and-future-conservation [Accessed 27 Aug. 2019].
6. https://www.ncbi.nlm.nih.gov/pubmed/29118219 [Accessed 27 Aug. 2019].
7. https://www.nature.com/news/1998/980924/full/news980924-5.html [Accessed 27 Aug. 2019].
8. https://sites.lsa.umich.edu/wp-content/uploads/sites/162/2016/07/Marshall-Wich-2016-Some-future-directions-for-primate-conservation-research-small.pdf [Accessed 27 Aug. 2019].
9. https://pdfs.semanticscholar.org/4d88/c340b73f47f35eda64f2b3c1166d468f1edb.pdf?_ga=2.4763952.285326929.1565721149-2084624878.1560189232 [Accessed 27 Aug. 2019].
10. https://www.weforum.org/agenda/2017/02/primates-are-closer-to-extinction-than-ever-says-new-study/ [Accessed 27 Aug. 2019].

INDEX

O

P

Q

R

S

T